Network Performance and Fault Analytics for LTE Wireless Service Providers

Deepak Kakadia · Jin Yang
Alexander Gilgur

Network Performance and Fault Analytics for LTE Wireless Service Providers

Springer

Deepak Kakadia
Network Analytics
Google
Mountain View, CA
USA

Alexander Gilgur
Technical Infrastructure
Facebook, Inc.
Menlo Park, CA
USA

Jin Yang
Wireless Technology and Strategy
Verizon Communications
Orinda, CA
USA

ISBN 978-81-322-3719-8 ISBN 978-81-322-3721-1 (eBook)
DOI 10.1007/978-81-322-3721-1

Library of Congress Control Number: 2017934864

Printed on acid-free paper

This Springer imprint is published by Springer Nature
The registered company is Springer (India) Pvt. Ltd.
The registered company address is: 7th Floor, Vijaya Building, 17 Barakhamba Road, New Delhi 110 001, India

Preface

This book describes network performance and fault analytics for LTE service providers from a practical perspective, which combines years of practical engineering experience working in the LTE service provider industry with data science and engineering insights. This unique combination of the three authors background provides innovative forward-looking approach to resolve long-standing complex network performance and fault management issues that span multiple domains.

Traditional telecommunication networks have been composed of separate vendor solutions from many domains. This resulted in disparate network management tools coming from domain-separated vendors. The radio access network (RAN) and the transport core network have been engineered and optimized in the past as separate and independent networks. This book proposes an end-to-end solution and hence an end-to-end network management and optimization architecture approach. The goal of an end-to-end optimization is a more consistent end-to-end user experience and higher network capacity and efficiency.

In this book, initial chapters present the fundamental building blocks from a bottom-up approach to provide enough background to understand the latter chapters which present solutions from a top-down approach. For example, we describe the main network types and characteristics of a typical LTE service provider network, which includes the RAN, Backhaul, Metro, and Core, as well as fundamentals of relevant data science techniques. These chapters equip the reader with the tools to understand the more advanced network performance and fault analytics methods and architectures which are presented in the latter chapters.

We would like to acknowledge many people who provided assistance on many network performance and fault-related projects at Verizon Communications, Google, and now Alphabet. At Verizon, Dr Jin Yang would like to thank Sanyogita Shamsunder, Bill Stone, Adam Koeppe, Tom Sawanobori, Ed Chan for their support on various projects, including radio network evolution, network planning, engineering and SON. Deepak would also like to thank Verizon for 10 wonderful years, in particular, Yong Gao, Tommy Broussard, Sundar Rangamani, Ashok Srivastava, Cindy Wells, Chris Neisinger for the outstanding support on various network-related projects ranging from the MPLS core, Metro, Backhaul, and network analytics. Deepak would also like to thank Google in the past year in particular Bikash Koley, Geng Lin, Mike Wiley, Dave Lefebvre, Ankur Jain, Kamran Sistanizadeh, and Matt Welsh for their trust, faith, and support. Authors would also like to thank their families for their understanding on the countless hours spent away from them on writing this book.

Mountain View, CA, USA — Deepak Kakadia
Orinda, CA, USA — Jin Yang
Menlo Park, CA, USA — Alexander Gilgur

About the Book

Network operators are faced with many new challenges, in particular, how to manage and operate a massively scaled network which consists of various network elements from different vendors and network domains. This book is intended to describe how to leverage emerging technologies Big Data Analytics NFV and SDN, to address challenges specific to LTE and IP network performance and fault management data in order to more efficiently manage and operate an LTE wireless network.

The proposed integrated solutions permit the LTE network service provider to operate entire integrated network, from RAN to Core, from UE to application service, as one unified system and correspondingly collect and align disparate key metrics and data, using an integrated and holistic approach to network analysis.

The LTE wireless network performance and fault consists of the network performance and management of network elements in EUTRAN, EPC, and IP transport components, not only individually but also inter-working of these components. The key metrics for EUTRAN include radio access network accessibility, retainability, integrity, availability, and mobility. The key metrics for EPC include MME accessibility, mobility and capacity, SGW, PGW capacity, and connectivity.

In the first parts of the book, we describe fundamental analytics techniques and various key network partitions—RAN, Backhaul, Metro, and Core of a typical LTE wireless service provider network. In the second part of this book, we develop more advanced analytic techniques that can be used to solve more complex wireless network problems. The second part of this book also describes practical and novel solutions for LTE service network performance and fault Management systems using Big Data Engineering. Self-organizing network architecture is presented as a way to utilize network performance and fault analytics to enable network automation. SON can significantly improve operational efficiencies and speed up the network deployment.

This book provides various ways to leverage data science to more intelligently and reliably to automate and manage a wireless network.

Contents

1 Network Performance and Fault Analytics for LTE Wireless Service Providers 1
1.1 Introduction 1
1.2 Motivation 1
1.3 Current Performance and Fault Management Architectures 8
1.4 Proposed Next-Generation Performance and Fault Management Architectures 9
1.5 Summary of Gaps in Current Network Performance and Fault Tools and Analytics 12
1.6 Book Outline 13

2 Analytics Fundamentals 15
2.1 Statistical Process Control 15
2.1.1 Central Limit Theorem 15
2.1.2 Applications of Central Limit Theorem: Bernoulli Trials 17
2.1.3 Examples of SPC for Bernoulli Trials 17
2.2 Outliers 19
2.2.1 QoS Outliers 20
2.2.2 Outliers: What Are They? 21
2.2.3 Outlier Detection: The Basic Approach 21
2.2.4 Advanced Methods of Outlier Detection 23
2.3 A Few Words About Queueing Systems 23
2.3.1 "True" Process Distributions for LTE Network Components 25
2.3.2 Little's Law: The "Big Three" of Queueing Dynamics 25
2.3.3 System Performance Laws 28
2.3.4 Conclusion 30
2.4 Forecasting 30
2.4.1 Time Series: Definition and Assumptions 31

2.4.2 Filling in the Gaps in Data . . . 32
2.4.3 Moving Averages . . . 34
2.4.4 EWMA . . . 35
2.4.5 EWMA Forecasting . . . 35
2.4.6 ARIMA Forecasting . . . 39
2.4.7 Selection of Forecasting Model . . . 39
2.5 Regression . . . 40
2.5.1 A Few Words on Terminology . . . 41
2.5.2 Linearizable Relationships . . . 41
2.5.3 The Main Idea Behind Regression . . . 43
2.5.4 Solving Eq. (2.5.2) . . . 44
2.5.5 Goodness of Fit . . . 45
2.5.6 Model Competition . . . 45
2.5.7 Analysis of Residuals . . . 47
2.5.8 Advanced Regression Methods . . . 50
2.5.9 Do We Have to Compete? . . . 53
2.6 Clustering . . . 54
2.6.1 Bundling the Curves . . . 55
2.6.2 Geographic Clustering . . . 58
2.6.3 Geographic Clustering of Signal . . . 58
2.7 Conclusion . . . 60
References . . . 60

3 Evolved Universal Terrestrial Radio Access Network (EUTRAN) . . . 61
3.1 RAN Architecture . . . 61
3.1.1 Evolved Node B . . . 62
3.1.2 Centralized RAN . . . 63
3.1.3 Home eNB . . . 64
3.2 Air Interface Specifications . . . 64
3.2.1 Radio Access Fundamentals . . . 64
3.2.2 Downlink Channel Structure . . . 66
3.2.3 Uplink Channel Structure . . . 66
3.2.4 LTE Mobility . . . 67
3.3 Evolved Node B—Base Station . . . 68
3.3.1 Baseband Processing . . . 69
3.3.2 LTE Scheduler . . . 70
3.4 RAN Network Management Model . . . 71
3.4.1 Performance Management . . . 72
3.4.2 Fault Management . . . 73
3.4.3 Configuration Management . . . 74
3.5 RAN Network Key Performance Indicators and Analyses . . . 75
3.5.1 Accessibility . . . 76
3.5.2 Retainability . . . 76

3.5.3 Integrity 77
3.5.4 Availability 78
3.5.5 Mobility 78
3.6 Summary 81

4 Enhanced Packet Core Network 83
4.1 EPC Architecture 83
4.1.1 SAE Gateways 85
4.1.2 Mobility Management Entity 86
4.1.3 Policy and Charging Resource Function (PCRF) 87
4.1.4 Home Subscription Server (HSS) 89
4.1.5 Application Services Domain (AS) 89
4.2 EPC Interfaces Specifications 90
4.2.1 S1-U Interface 90
4.2.2 S1-C Interface 91
4.2.3 S5/S8 and SGi Interfaces 92
4.2.4 Gx/Gxc and Rx Interfaces 93
4.3 EPC Network Management Model 94
4.3.1 MME Mobility Measurements 94
4.3.2 MME Session and Subscriber Management 96
4.3.3 SGW Measurements 97
4.3.4 PDN-GW Management 98
4.3.5 PCRF-Related Measurement 98
4.4 EPC Key Performance Indicators and Analyses 99
4.4.1 MME Accessibility, Mobility, and Capacity 99
4.4.2 SGW and PGW Capacity and Connectivity 99
4.4.3 Service-Specific Performance and Analyses 100
4.5 Summary 102

5 Overview of IP Transport Network Architectures—Backhaul, Metro, and Core Network 103
5.1 Ethernet Backhaul Network Architectures 105
5.1.1 Ethernet Network Analytics—Optimal Hub Sites 108
5.1.2 Ethernet Backhaul Network Analytics—Measuring Actual Customer Experience 109
5.1.3 Backhaul Analytics Insights 111
5.2 Core and Metro Network Architectures 113
5.2.1 Core and Metro Network Elements—ROADMS 120
5.2.2 Core and Metro Network Elements—IP/MPLS Routers . . . 121
5.2.3 Core and Metro Transport Links 121
5.2.4 Core and Metro Network Analytics and SDN 122
5.2.5 Multilayer Network Performance Correlation 122
5.3 TCP Analytics 129
5.4 QoS Analytics 131

5.5 Network Analytics and SDN . . . 134
5.6 Summary . . . 135

6 **Advanced Analytics** . . . 137
6.1 Statistical Process Control for E-UTRAN KPIs . . . 137
6.1.1 Retainability . . . 137
6.1.2 The KPIs: UL and DL Throughputs . . . 142
6.1.3 The KPIs: UL and DL Latencies . . . 144
6.1.4 Nonparametric SPC—A Cleaner Way to Show the UCL and LCL of the Non-normal Variable . . . 145
6.1.5 The KPIs: Availability . . . 147
6.1.6 The KPIs: Mobility . . . 147
6.2 Real-World Outliers . . . 148
6.2.1 Introduction . . . 148
6.2.2 Advantages of Tukey's Method . . . 151
6.2.3 Disadvantages . . . 151
6.2.4 Practical Use Cases . . . 151
6.2.5 Aggregating Data . . . 155
6.3 Queueing and LTE Data . . . 157
6.3.1 User Throughput . . . 157
6.3.2 Effects of Data Transformations on Distributions . . . 158
6.3.3 Retransmissions . . . 160
6.3.4 Router Load Balancing/Scheduling . . . 162
6.4 Conclusion . . . 164
References . . . 165

7 **Next Generation Network Performance and Fault Analytics Management Systems** . . . 167
7.1 Overview . . . 167
7.1.1 Network Measurements . . . 168
7.1.2 Standards . . . 170
7.2 Network Performance and Fault Management Systems Architectures . . . 171
7.2.1 Incremental Migration to Next-Generation Performance and Fault Analytics Platform—Integration of Existing NMS Systems at Top Layer . . . 172
7.3 Design Considerations for Carrier Grade Network Analytics System Optimized for Performance and Cost . . . 175
7.3.1 Carrier Grade Network Performance and Fault Analytics System Features and Functionalities . . . 176
7.3.2 TS 32.401—Performance Management Concept and Requirements . . . 177
7.4 Network Traffic Analytics Architectures . . . 178
7.5 Network Analytics Engine . . . 179

7.5.1 Current Approach 180
7.5.2 Limitations of Current Approach 180
7.5.3 Proposed Approach 181
7.6 Summary 182

8 Self Organizing Networks (SON) 183
8.1 SON Introduction 184
8.1.1 Self-configuration 184
8.1.2 Self-optimization 185
8.1.3 Self-healing 186
8.2 SON Architecture 186
8.2.1 eNB SON Functionalities 187
8.2.2 Centralized SON Functionalities 188
8.2.3 Hybrid SON Control Mechanism 189
8.3 Self-optimization Features 189
8.3.1 Mobility Robustness Optimization 190
8.3.2 Mobility Load Balancing 192
8.3.3 Coverage and Capacity Optimization 193
8.4 SON Deployment and Performance Benefits 194
8.4.1 SON Deployment 195
8.4.2 Performance Benefits 198
8.5 Evolved Network Architecture 199
8.6 Summary 200

9 Summary 201
9.1 LTE Wireless Network Element 201
9.2 Analytics 202
9.3 Solutions 202
References 203

About the Authors

Deepak Kakadia is currently at Google in MountainView CA, working in the area of NFV, SDN Network and Analytics. Previously, from January 2013 to January 2015, he was the team leader, Distinguished Member of Technical Staff (DMTS), IP network architect, with Verizon Labs, leading Network QoS Analytics and Network QoS Optimization for LTE Wireless Network Service Provider Networks in Palo Alto, CA. From May 2005 to January 2013, he was with Verizon/Verizon Wireless, in the Head Quarters Network Planning Group in Walnut Creek, California, USA, since. Previously he was a staff engineer, IP network architect at Sun Microsystems Inc., Menlo Park, California, for a total of 11 years since 1994. He also worked at Corona Networks as a principal engineer in the Network Management Systems group; Digital Equipment Corp, where he worked on DEC OSF/1; and Nortel Networks (Bell Northern Research) in Ottawa, Canada. He received a certificate in networking from the Department of Electrical Engineering at Stanford University, Palo Alto, CA. He received a Bachelor of Engineering in Computer Systems, a Master of Science in Computer Science, and has completed Ph.D. qualifiers and course work in Computer Science. Deepak is a recognized industry expert in networking architecture and analysis, invited speaker at many international conferences, where he was approached by Springer to write a book on this subject. He has authored at time of writing 60 awarded patents and filed over 80 patents in the area of network and systems management and wireless technologies.

Jin Yang received the B.Sc. (Honors) and Ph.D. degrees from Tsinghua University. Dr. Jin Yang is a fellow at Verizon Communications, responsible for wireless technology and strategy. She is leading next-generation wireless network architecture and technologies, including self-organizing network, machine-type communications, heterogeneous network, LTE-advanced and 5G access technologies. She has played a key role in the development and commercialization of LTE network in 2010, Verizon's choice of LTE as 4G technology in 2007, various CDMA network developments since 1995 at Verizon, Vodafone and AirTouch Communications. She has managed the development of CDMA design and planning tool at AirTouch. Dr. Jin Yang was an adjunct professor at Portland State

University where she taught wireless communications. She has contributed to establish US IS-95 CDMA standards. Dr. Jin yang has more than 35 granted patents and 30 pending patents on wireless communications. She has published numerous papers and is the co-author of 3 published books.

Alexander Gilgur is a data scientist and systems analyst with over 20 years of experience in a wide variety of domains—control systems, chemical industry, aviation, semiconductor manufacturing, information technologies, and networking. He received his M.S. in electrical engineering from Moscow Institute of Chemical Engineering and M.S. in sport psychology from Capella University. He has completed Six Sigma Black Belt certification requirements and holds a professional engineering license for chemical engineering. Alexander has authored and co-authored a number of know-hows and published numerous papers and patents. He has a solid track record of implementing his innovations in production and enjoys applying the beauty of mathematics and statistics to solving system capacity and performance problems and is interested in non-stationary processes, which make the core of IT problems today. Presently, he is a network data scientist at Facebook. Prior to Facebook, he was a network analyst at Google and a faculty member/lecturer for machine learning at UC Berkeley's MIDS program. He is also a father and a husband, a skier, a soccer coach, and a music aficionado. Alex's technical blog is at http://alexonsimanddata.blogspot.com.

Abbreviations

DL	Downlink (base station to user device)
eNB	Evolved Node B
EPS	Evolved Packet System
E-UTRAN	Evolved UMTS Terrestrial Radio Access Network
HARQ	Hybrid Automatic Repeat Request
HO	Handover
HRPD	High Rate Packet Data, referred to 1xEV-DO
HSPA	High-Speed Packet Access, referred to HSDPA/HSUPA
IOS	Inter-Operability Specification,
IP-CAN	IP Connectivity Access Network
LTE	Long-Term Evolution
MAC	Media Access Control
MBSFN	Multicast Broadcast Single Frequency Network
MME	Mobility Management Entity
MSC	Mobile Switching Center
PCRF	Policy and Charging Rules Function
PDCP	Packet Data Convergence Protocol
PDN GW	Packet Data Network Gateway
PSTN	Public Telephone Switching Network
QoS	Quality of Service
RAI	Routing Area Identification
RAN	Radio Access Network
RLC	Radio Link Control
RNC	Radio Network Control
RRC	Radio Resource Control
SAE	System Architecture Evolution
SGW	Serving Gateway
SQL	Structured Query Language
UE	User Equipment
UL	Uplink

UMTS	Universal Mobile Telecommunications System
USIM	Universal Subscriber Identity Module
UTRA	UMTS Terrestrial Radio Access
UTRAN	UMTS Terrestrial Radio Access Network
VCC	Voice Call Continuity

Chapter 1
Network Performance and Fault Analytics for LTE Wireless Service Providers

1.1 Introduction

LTE (long-term evolution) commercial carrier trials started around 2009–2010 in scattered locations around the world by wireless service providers such as TeliaSonera, Telecom Italia, Telstra, Verizon Wireless, and AT&T whose successes in superior performance resulted in production widespread deployments rapidly. LTE has proven to be an open high-performance wireless network access technology to provide massive data rates and efficient use of spectrum with commercial benefits that motivated both CDMA-based carriers and GSM-based carriers to put LTE migration as the main wireless access technology for most carriers around the world and equally devise manufacturers.

LTE wireless carriers typically spend approximately 25% of their expense budget on operational costs to maintain and manage the network. LTE poses new operational challenges including managing the massive amounts of network performance and fault data and correlating the radio access network (RAN) with the IP network. RAN network management systems typically do not integrate with IP network management systems, which is required to correlate data to provide the operator with an end-to-end visibility and insight to the operational aspects of the network, critical to the optimal management of the LTE service production network.

1.2 Motivation

LTE wireless service networks have migrated from a predominately legacy circuit-switched paradigm to an IP paradigm. The legacy telecom circuit switch had matured fault and performance tools developed by the same vendors who manufactured the Network Equipment and has served the industry well over the years. However, LTE wireless service networks have three fundamental problems as follows:

D. Kakadia et al., *Network Performance and Fault Analytics for LTE Wireless Service Providers*, DOI 10.1007/978-81-322-3721-1_1

1. The LTE equipment deployed in most typical environments is from different vendors. Even though standards such as 3GPP define interfaces to ensure interoperability, there are always integration issues. It is very difficult for one vendor to provide an end-to-end performance and fault tool which includes support for a competing vendor.
2. Most of the current available vendor tools only focus on their own network element KPIs and internal implementation, which often is of no value to the operator. The equipment vendor has no interest in providing visibility to the operator on how well the tool is performing as it can be used to compare with other vendors. Vendors have a vested interest in providing the absolute minimal visibility of their device which may expose limitations, such as queue backlogs due to a poor QoS scheduling algorithm.
3. Service providers typically view OAM as an afterthought, often with poorly qualified staff who rely on the vendor to provide a tool to manage. For example, the vendors the PGW have countless KPIs, which are mostly useless to the LTE operator, but provide implementation-specific data useful to only the vendor, often not even implemented properly. Vendors typically want to sell their own management system who only they understand the KPIs and can provide some insights that may or may not be useful to the operator.

This book has been written to provide LTE wireless service operators a novel approach to performance and fault management, where useful information can be pulled from the LTE network and make a difference to the business. The proposed approach takes a very simple route:

1. LTE Network Specifications:
   (a) Define the dynamic behavior of the LTE network map to a language that can produce a specification which can be understood by a computing machine, such as finite-state machines, Petri nets, and communicating sequential processes (CSP).
   (b) Define specifications of timings for these dynamic behaviors. This will be best accomplished by using a dynamic modeling language that can support the notion of time-outs, such as timed CSP.
2. Measure actual LTE network behaviors and timings—Measure the actual dynamic behaviors in the network. This will require some transformations from raw network measurements into the model that was used to define the dynamic behavior. Any deviations will be able to immediately identify root cause of network faults. Performance can be analyzed using simple queueing theory.
3. Analyze the measured results against specifications to determine corrective action. This process will allow the operator to automate the processing of massive data to draw useful conclusions with specific goals, such as fix faults, improve performance by adjusting QoS parameters.

The current approach to network performance and fault management is to take counters and display to the user. The proposed approach is to define models and

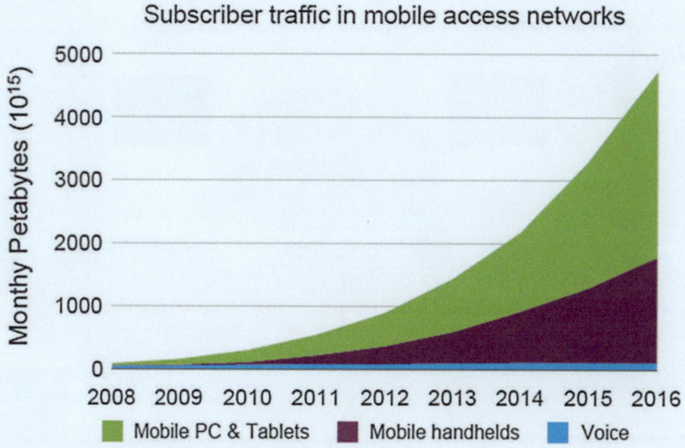

Fig. 1.1 Traffic growth in mobile access network

specifications, measure, compare against specifications, and draw useful insights and corrective action to make improvements in the LTE network.

Wireless networks and services are becoming more and more complex. This is driven by exponential traffic growth, network densification, and varieties of services. Subscriber traffic in mobile access networks has increased significantly with rapid penetration of smartphone and tablets, as illustrated in Fig. 1.1. The trend is going even higher with the Internet of Things (IoT) and Internet of Everything (IoE). Wireless service providers are evolving network management system toward network performance and fault analytics-based architecture. This will enable automatic network management, maintain superior performance, increase revenue, and decrease costs.

Self-organizing network (SON) and software-defined networks (SDN) have gain attraction recently; service providers are striving to utilize analytics to further improve efficiency of network automation, flexibility, and adaptation. Network performance and fault analytics have proven valuable means to generate insights on customer behaviors and retarget advertising. However, coordinating and generating insights on network traffic, equipment parameters, and network optimization required significant domain knowledge and evolution of traditional practice of network management and optimization. In LTE environments, this domain knowledge spans IP, RAN, optical, MPLS, QoS as well as operational work flows, which can take years of experience and different skill sets. This book attempts to bridge this gap.

Figures 1.2 and 1.3 show a diagram of a typical LTE service provider network and network elements that are monitored. The diagram shows the core network which has metro rings connected that geographically span a few kilometers for dense markets or hundreds of kilometers for sparse markets such as rural area. Within a metro area, there are mobile switching centers that simply aggregate many cell sites that contain the eNodeB, which is the demarcation point between the radio

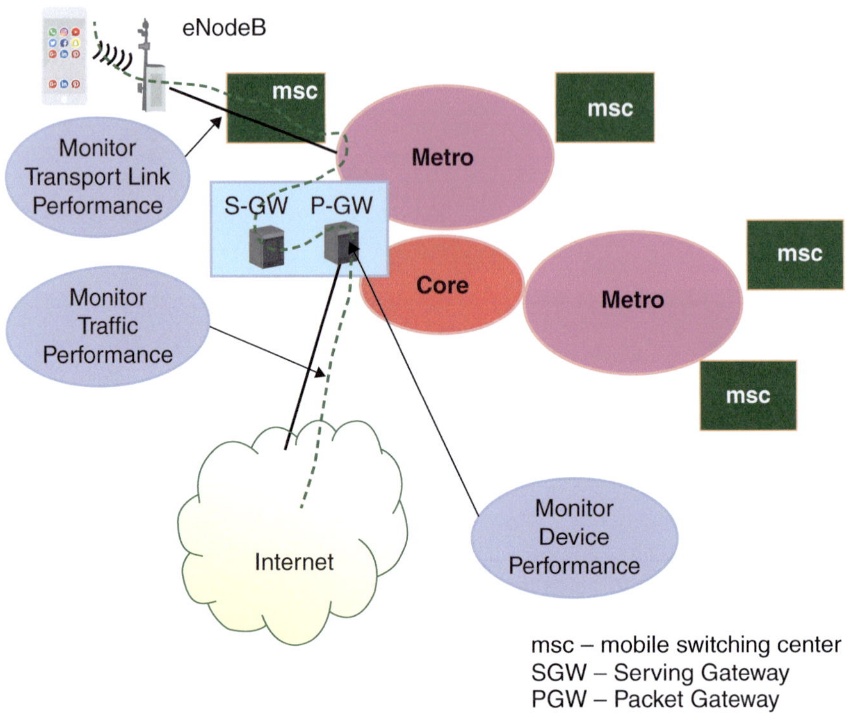

Fig. 1.2 Typical LTE service provider network topology and monitored elements

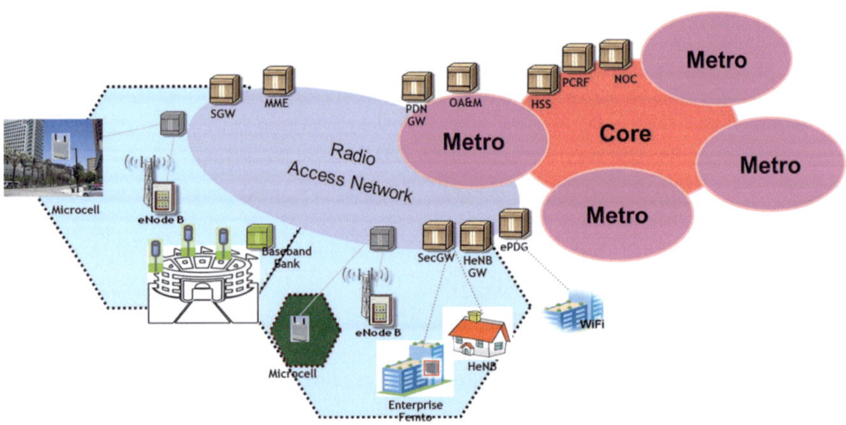

Fig. 1.3 LTE carrier network expanded to show macro-, micro-, pico-, and femto-cell examples

access network (RAN) and the Ethernet Backhaul network, which connects to the packet core (consisting of the SGW and PGW). This book will concentrate on data plane user traffic, which is why control plane nodes such as the MME are not shown. The following describes in further detail the elements within the LTE architecture that are monitored for performance and faults:

1. Network Elements—This class of elements include physical boxes that process traffic and forward onto next hops. Examples and categories include the following:

   (a) Various radio nodes—macro-, micro-, pico-, femto-, and Wi-fi nodes.
   (b) Layer 2 Network Elements—LTE SGW/MME, Layer 2 switches, firewalls, and NIDs.
   (c) Layer 3 Network Elements—Layer 3 routers, LTE PGW, HeNB GW, ePDG, security GW, firewalls, load balancers, and NAT devices.

   Tools to monitor network elements fall into two categories as follows:

   (a) SNMP—These include performance management and fault management devices that continuously poll the device to ingest counters, or receive traps when the device itself detects a fault and notifies the management system. This is typical of IP devices.
   (b) Element Management Systems (EMSs)—These are typically vendor-specific boxes, manufactured by the same vendor that manufactured the physical device and often ingests proprietary data from the NE to the EMSs but advertises standards-based open northbound interface. Vendors typically overcharge for this system as there is lack of competition.

2. Transport Links—This class of element includes devices that inject synthetic packets into the network and measure the time it takes to go from one box to another box. These devices attempt to measure the average latency of particular links. These devices are plagued with key limitations as follows:

   (a) As the number of links increases, the number of boxes increases, both growing exponentially.
   (b) As the required accuracy increases, the frequency of packet injection must increase, compromising customer traffic congestion as the synthetic packet starts to consume the same bandwidth capacity as revenue-generating consumer traffic.
   (c) As the required accuracy increases on a path that itself is composed of many segments, the number of boxes required increases to be installed at each segment.
   (d) This approach does not work on LTE devices such as SGW and PGW due to the fact the synthetic packet will not be forwarded across the SGW or PGW, so there is no way to measure the latency across these LTE components.

3. Traffic—The main tool to measure and monitor traffic is packet sniffers. This class of tool requires manual efforts to sniff links and examine manually the

traffic for problems. Due to the advances in link speeds, the manual process of sifting through massive amounts of traffic data is no longer a practical approach.
4. Devices—Subscriber devices are used to monitor and report access network performance, as well as application performance on a periodic basis and/or events driven. Intensive benchmark tests and third-party tests are used to provide both network status as well as user experience.

As we are evolving LTE access network from a macro-cellular network to a heterogeneous network involved macro-, micro-, pico-, and femto-, as well as integrated Wi-fi network, clear signs of stress are becoming increasingly apparent. Segregated and point solutions require a lot of manual efforts and expertise to correlate issues.

Wireless LTE service providers are struggling to manage and optimize their networks. One of the primary reasons for this problem stems from the belief that management tools are an afterthought, the primary focus of being the evaluation and deployment of the Network Equipment and was expected that equipment vendors would provide adequate tools to manage their equipment. The disconnect lies in the fact that service providers have now become the integrators and to integrate the various different vendor network devices is disproportionately more complex than managing individual elements. The problem is further compounded by the change from the telecommunications network-based network management tools to IP-based network management and the completely different skill sets required to effectively managing the network. Traditional telecommunication's vendors such as Nortel and Lucent manufactured all the equipment from base stations, RNCs, even Class 5 telecom switches, as well as the tools to manage these devices. The historical circuit switched based model has evolved from a single vendor-based solution to a mixed-and-matched vendors-based model. The network management solution vendors tried to address the need for managing IP-based service provider networks by recycling and retrofitting legacy enterprise-designed caliber network management systems.

The sheer amount of telemetry network monitoring data is orders of magnitude more than what the legacy IP enterprise network management systems were designed. The fundamental limitation is not only the data store but also the methodology. Classical enterprise IP network management systems collect all the data and then store in a relational database. This model does not scale for two reasons:

1. As the amount of data stored in relational databases increases, to perform JOIN operations increases exponentially. We saw cases where we can run only 20 SQL queries per night since each query took hours to execute.
2. The amount of manual efforts required to sift through massive amounts of LTE data has increased several orders of magnitude. Typical enterprise networks involved a few hundred routers for a large Fortune 500 company each with less than a hundred KPIs or MIB counters. LTE service providers now manage thousands or hundred thousand devices, some devices having thousands of variables.

Most commercially available performance management solutions have the same fundamental architectural model and limitation. Current network management or analytics tools are focusing on element scalar counters or key performance indicators (KPIs) where increasingly massive amounts of simple numbers are collected and stored using an outdated relational database model. The data then is retrieved using SQL queries and JOIN operations which do not scale. Further data storage requirements exceed feasible business budgets, and time-intensive human effort and skill are often required to extract meaningful insights, after a problem has occurred.

This book proposes a novel approach, where data is immediately analyzed and decided upon based on pre-programmed analytics. This has profound advantages as follows:

1. Immediate and predictive insights based on deterministic events that violate specifications, opportunities now exist to take corrective action before customers are impacted.
2. Reduction in human manual efforts sifting through massive data, resulting in automation and increased operational efficiencies.
3. Reduction in storage requirement resulting in massive scale as data rates and volumes increase.

The novel approach essentially starts from addressing the key problem areas and then working on building an optimal solution for solving that set of particular problems in an optimal manner—specifically designed for LTE service network, following these tenets:

- At the end of the day, what matters above all is the customer experience. Start with quantifying the customer quality of experience.
- Collect all necessary data, analyze, and then keep only useful or suspicious data. This is the focus of this book. It makes no sense to collect massive amounts of data and expect humans to sift through all this data manually. Storage costs, latencies, and labor costs make this infeasible. Network analytics is about analyzing the data as it is streaming in, in an automated fashion, and at the speed of computers. You have all the information you need to make a deterministic decision for 90% of all cases at this point of time. You collect data from various sources, correlate, analyze, and provide an answer to the network engineer for tactical objectives and provide an answer for capacity planning engineer for long-term strategic objectives. Further, this same data processed different ways can be forwarded to the marketing analytics team for correlation with other marketing-related data.
- Simplification and Cost Reduction. The proposed approach will be able to immediately identify the health of the subscriber's traffic, the health of the transport network, and only start to collect data needed to proactively investigate suspicious areas, requiring more granular KPIs to be collected adaptively and analyzed as needed. This approach will eliminate the need of many legacy single-purpose tools. For example, if the traffic is monitored on all links, on a millisecond basis, then there is no need for legacy tools that compute bandwidth

based on octet counts over 5-min intervals which is less accurate and does not expose any intermittent congestion on a millisecond basis. Further, there is no need for synthetic packet latencies that estimate latency on round-trip or one-way basis for that synthetic packet not the actual customer traffic. Some LTE devices such as the SGW and PGW transform the packet, so these synthetic tools will never be able to measure these latencies as proposed in this book. Finally, the storage requirements will be dramatically less, saving tens of millions of dollars by not storing data that very few people use. For some use cases, low-cost Hadoop-based systems make sense and that will be preprocessed, compressed data for not only less storage but also faster analytics.

We will outline various network elements and corresponding management system in individual chapters and provide an end-to-end optimized solution in late chapters with some practical examples.

1.3 Current Performance and Fault Management Architectures

As LTE service providers moved from mature legacy circuit performance and fault management tools to IP performance management tools, the vendor community essential re purposed enterprise grade network performance and fault management

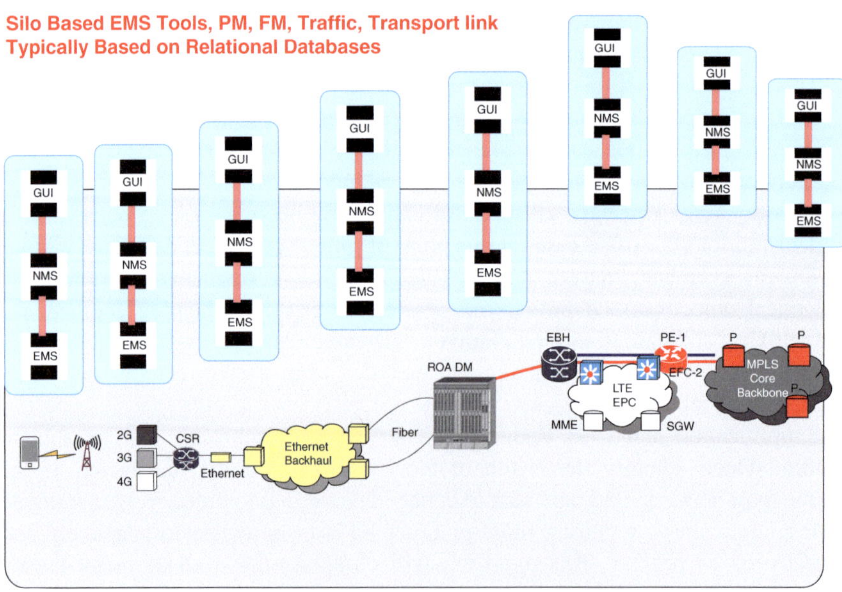

Fig. 1.4 Current legacy performance and fault management systems

tools. The problem was scale- and integration-specific to LTE environments. Figure 1.4 shows a typical performance and fault management system deployment typically found in many LTE service provider environments. As can be seen most tools are silo-purposed tools, yet there is tremendous overlap. Most EMSs have their own limited implemented performance and fault management system, and there is additionally a best of breed companywide performance and fault management tool, creating redundant functions. There is poor integration, and manual efforts are required to correlate IP and LTE network issues.

Current network management includes tools specifically designed for network operational performance and configuration management, device and network real-time traffic analyses, RF design and optimization, and capital planning and forecasting. There are also reports for end-to-end service performance, such as data, voice, E911, and SMS. Service providers measure performance of various interfaces in the network, track backhaul, and security in the network. Very specific tools are also used for IP address, physical cell identification (PCI), and neighbor relation managements.

Network operators are consistently struggling with inter-working of various network elements and tools to manage those network elements. Proprietary interfaces among various network elements prevented speedy end-to-end network performance and fault analytics.

1.4 Proposed Next-Generation Performance and Fault Management Architectures

Real-time performance and fault management is the foundation of a smart network, a network that can be redefined and optimized for various services using varieties of network elements. The performance and fault analytics can take inputs from user device, radio access nodes, session management, and various applications and policies, as illustrated in Fig. 1.4 It then makes intelligent decision to feed into policy and resource management to drive configuration and parameters of various network elements (Figure 1.5).

Figure 1.6 shows an example of the proposed integrated performance and fault analytics management framework, which removes redundancy and additionally adds missing measurements such as traffic, which is the nucleus of the system. As can be seen, the transport link data is not shown but will be derived from the network element data, in particular physical interfaces. There are 2 NE data shown, one is LTE and other is IP. LTE performance and fault data is usually proprietary vendor-specific format and often in bulk files, whereas NE IP data is usually based on SNMP. Finally, the biggest difference is the traffic data collected, which is the real innovation. It is the traffic data from which the analytics is based on. The traffic data is measured on a per-segment per-QoS basis and compared against specifications. Each QoS in LTE traffic has specific packet delays and loss rates. The

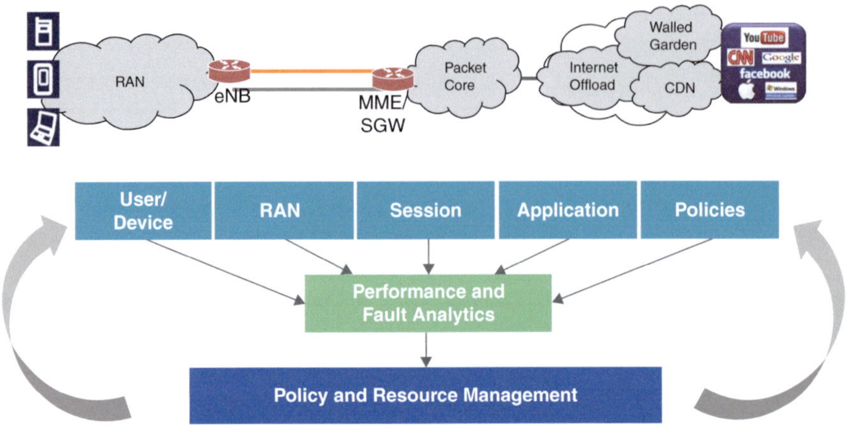

Fig. 1.5 Four smart networks

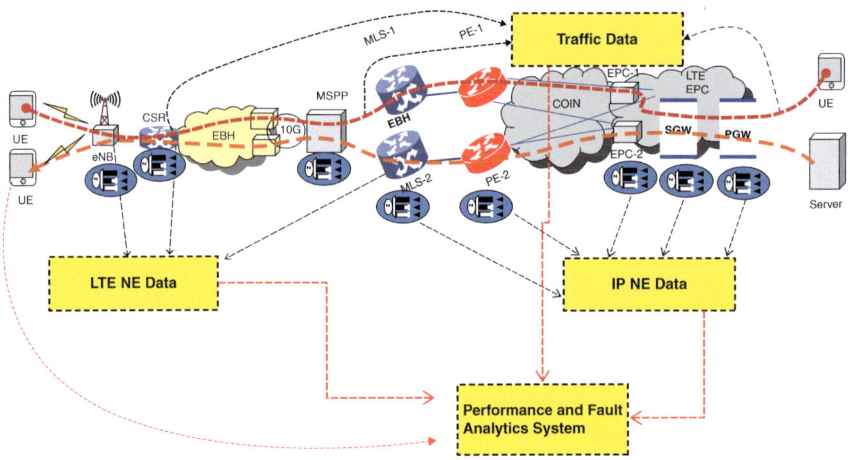

Fig. 1.6 Proposed performance and fault analytics in LTE network infrastructure (please add a link from UE)

traffic management system will monitor the traffic and as soon as a problem is detected, other sub-systems will be alerted to determine root cause, which may invoke other more granular traffic measurements. If the traffic is running smoothly, there is no point in collect vast amounts of network element data.

Figure 1.7 shows a clearer diagram of the main components of the proposed architecture. As can be seen, all data flows into the analytics engine, and it is proposed to integrate the performance and fault analytics because performance data can predict faults before customers are impacted.

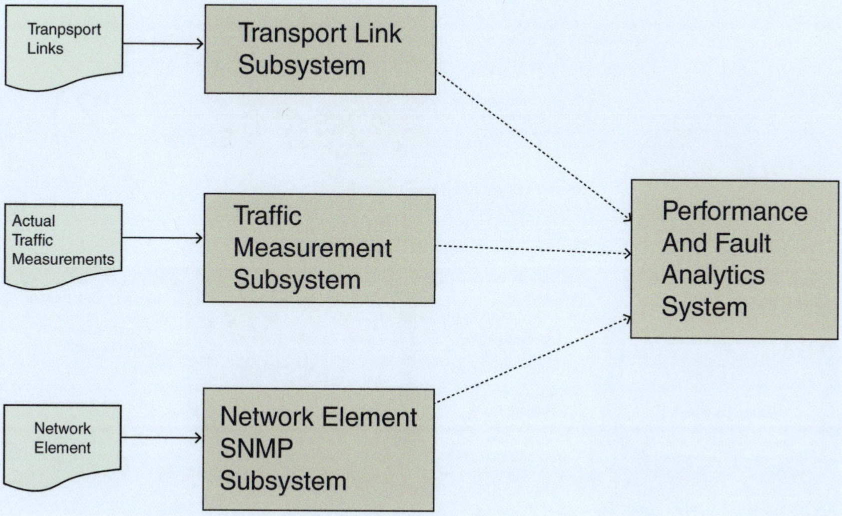

Fig. 1.7 Software abstract architecture of performance and fault analytics. (replaced with hybrid SON architecture?)

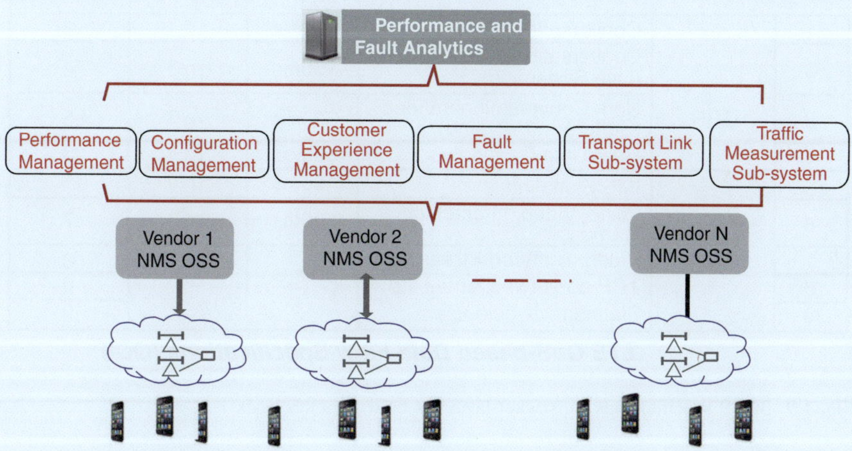

Figure 1.8 shows one example of the data flows in the proposed network analytics framework to show how this system will work. The traffic that is measured at various key segments will show the actual latencies, throughput, and error rate experienced by subscribers. Flows associated each QoS value have a specified target-budgeted latencies on each segment, which is based on 3GPP specifications as shown in Fig. 1.8. Network resource management could be adjusted according to analytic engine resided in a centralized location (Fig. 1.9).

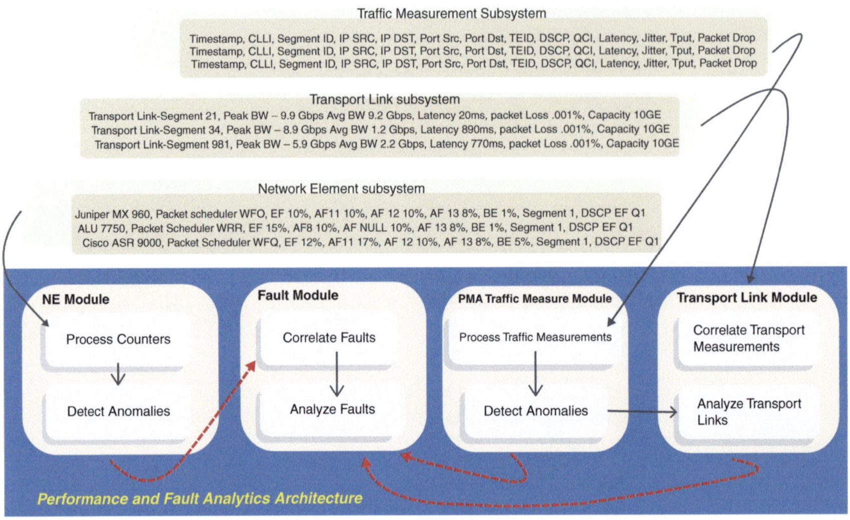

Fig. 1.8 Detailed data flows in analytics framework—an example

QCI	Bearer Type	Application Example	Packet Delay	Packet Loss	Priority
1	GBR	Conversational VoIP	100ms	10^{-2}	2
2		Conversational Video (Live Streaming)	150ms	10^{-3}	4
3		Non–Conversational Video (Buffered Streaming)	300ms	10^{-6}	5
4		Real Time Gaming	50ms	10^{-3}	3
5	NON–GBR	IMS Signaling	100ms	10^{-6}	1
6		Voice, Video, Interactive Games	100ms	10^{-3}	7
7		Video (Buffered Streaming)	300ms	10^{-6}	6
8		TCP apps (web, email, ftp)			8
9		Platinum vs. gold user			9

LTE QoS-based Data Flow Specifications (QCI)

Fig. 1.9 3GPP specifications for packet latencies on a per-QoS basis

1.5 Summary of Gaps in Current Network Performance and Fault Tools and Analytics

The current network performance and fault management tools are typically very basis counters and little or no insights or correlation across domains. LTE service providers have vast amounts of complex data, and manual efforts for technicians to reactively solve performance or faults are neither scalable nor sustainable. Tremendous amounts of data is in fact collected and stored in a relational database;

however, very little of this data is actually used to really solve problems. This problem is expected to get worse overtime as LTE data is increasing exponentially and pressure to reduce OPEX is increasing.

This book proposes an alternate approach to network and fault analytics starting from the ground up. The most important measure is to verify that subscribers are experiencing the best possible quality of experience (QoE) based on the QoS provisioned and measurements start at the traffic on a per-QoS and per-segment basis. Only when problems are detected at the traffic layer, only relevant and more granular data is collected needed to determine root cause. This is done in an automated manner. The network performance and fault analytics engine will proactively detect potential issues and deterministically determine root cause. The net effect is to identify issues before the customer is impacted and take corrective action. The secondary benefit is that only data that is useful to the operator and prepared for easy understanding is stored. Most of the data that is collected is processed and analyzed in stream and then dropped. The answer is given to the operator and related evidence. This is in contrast to current legacy systems that store vast amounts of data requiring manual efforts to infer insights. The proposed approach pre-programs the intelligence to process and interpret data and provide a useful insight in an automated manner to the technician. This approach dramatically improves customer QoE, time to resolve a problem, the amount of man hours to resolve a problem and automate many manual tasks needed to correlate and determine root cause of network faults. Further, the system is designed to be proactive; as soon as latencies increase, the system knows something is wrong, there is enough information to predict a fault will occur, and the system can automatically pinpoint source of issue.

1.6 Book Outline

The LTE service provider network is composed of multiple network partitions such as RAN, EPC, and Core each with unique elements and architecture. This book will describe common components and then drill down into each network component.

Chapter 2—Analytics Fundamentals: This chapter will provide the reader with a review of fundamental analytics such as statistical process control, outliers, queueing theory, forecasting, regression, and clustering. These fundamental concepts will be applied to more complex LTE radio and core network performance analysis in Chap. 6, which will also offer more advanced techniques, building on the fundamentals learned in this chapter.

Chapter 3—Evolved Universal Terrestrial Radio Access Network (EUTRAN)—This chapter will provide the reader with an in-depth explanation of the radio network architecture, basic elements, network management, and performance and fault analysis.

Chapter 4—Enhanced Packet Core Network (EPC)—This chapter will provide the reader with an explanation of the EPC architecture, network elements, network management, and performance and fault analysis.

Chapter 5—Overview of IP Transport Network Architectures—Backhaul, Metro, and Core Network—This chapter will provide the reader with an explanation of the main components of the LTE service providers' IP network, which include the backhaul, that connects the cell towers to the service providers' mobile switching center (MSC). This is then followed by the Metro network, which interconnect the MSC in a particular region and can scale out in a hierarchical architecture to connect multiple metros to a regional network. The Core network, also referred to as the backbone, interconnects the LTE service providers Metro networks which allow any to any connectivity for the PDN, including access to the Internet.

Chapter 6—Advanced Analytics—This chapter will build on the theoretical analytic concepts of Chap. 2 to show how these techniques can be used to provide new insights in real LTE service network performance and fault analysis. This chapter will provide the reader with a deeper explanation of how analytics concepts can be applied to more complex LTE service provider network performance and fault operations to predict and resolve issues that traditionally were done manually.

Chapter 7—Next Generation Network Performance and Fault Analytics Management Systems. This chapter will provide an explanation of the limitations of current management systems available to LTE service providers and propose new concepts to more efficiently manage LTE service provider networks in the area of performance and fault analysis. The concepts learned in the previous chapters in understanding the network components and analytics will all come together in this chapter.

Chapter 8—Self Organizing Networks (SON)—This chapter is another application of LTE network analytics that enables LTE service providers increased operational efficiencies by self-configuration, self-optimization, and self-healing.

Chapter 9—Summary—This chapter summarizes how current performance and fault management systems can be significantly improved in terms of operational efficiencies and subscriber quality of experience by leveraging new advances in big data analytics specifically for LTE service providers.

Chapter 2
Analytics Fundamentals

In this chapter, we are introducing the key methods of data analytics and describe the underlying principles, assumptions, and the thought process behind applying these methods in LTE performance and fault analysis. We start with the most basic form of time-series analysis (statistical process control) and progress through the mathematical concepts of outliers, queueing theory, system performance laws, forecasting, regression, and clustering.

2.1 Statistical Process Control

Statistical process control (SPC) today is the hallmark of a mature technology. When properly established, not only does it allow NOC operators to promptly detect, quantify, scope, and analyze root causes of problems in the network, leading to quickly solving the problems; it also makes forecasting easier and more intuitive.

A number of SPC techniques have been developed since the approach was first introduced by Walter Shewhart of Bell Laboratories in the early 1920s. Shewhart was one of the "early rejecters" of the approach to production process analysis from the standpoint of "normal," or Gaussian, distribution. We have to keep in mind that in the 1920s, the Central Limit Theorem (CLT) was not yet as firmly established in the world of probability and statistics as it is today.

2.1.1 Central Limit Theorem

Central Limit Theorem today serves as a convenient justification for treating any random process as normally distributed, which ensures simplicity of mathematical

D. Kakadia et al., *Network Performance and Fault Analytics for LTE Wireless Service Providers*, DOI 10.1007/978-81-322-3721-1_2

analysis of data leading to actionable results. Indeed, if we assume normal distribution of data, we can get away with using well-established simple methods developed by William Gossett, Karl Pearson, and Ronald Fisher for statistical hypothesis testing. CLT gives us a way to justify such assumption: If we have "enough" independent samples that are "big enough," then the means (arithmetic averages) of these samples converge to a normal distribution with the mean equal to the "mean of the means" and the variance computed as the sum of variances of the samples.

$$\lim_{N\to\infty} \frac{\sum_{i=1}^{N} \mu_i}{N} = \mathbb{N}(\mu, \sigma) \tag{2.1.1}$$

Here

$\mathbb{N}(\mu, \sigma) =$ normal distribution with

mean $= \frac{\sum_{i=1}^{N} \mu_i}{N}$ and standard deviation $= \sqrt{\sum_{i=1}^{N} \sigma_i^2}$

In Eq. (2.1.1),

$$\text{mean} = \frac{\sum_{i=1}^{N} \mu_i}{N} \tag{2.1.2.a}$$

$$\text{standard deviation} = \sqrt{\sum_{i=1}^{N} \sigma_i^2} \tag{2.1.2.b}$$

Lyapunov's equation for the variance of the aggregate of independent samples (sum of squares of variances) marked an important milestone in probability theory and statistics: It allowed treating samples as dimensions of the process being analyzed, as long as they were independent.

This simple construct enabled a number of breakthroughs in data science; most importantly, it allowed engineers to quickly develop a number of statistically sound robust "rule of thumb," such as Western Electric Rules and Nelson Rules, for statistical process control. It also allowed a delineation of long-term and short-term process capability that has become one of the cornerstones of the Six Sigma Methodology, which became the next breakthrough in statistical quality control developed in the late 1980s and early 1990s and dovetailed into the concept of Design for Six Sigma (DFSS), which is currently used by manufacturers of hi-tech equipment all over the world.

2.1.2 *Applications of Central Limit Theorem: Bernoulli Trials*

Disclaimer: All charts and data in this chapter are produced by simulation and do not reflect actual vendor data.

"Bernoulli trials" is a term in probability theory representing a scenario where a definition of insanity falsely attributed to Einstein ("doing the same thing over and over and expecting different results") is challenged: Given a binary random variable X (or a process that can be in one of two states, typically "success" and "failure," or 0 and 1), we repeat the same action that produces and tallies the number of trials and the number of times when $X = 2.1$. Then,

$$\Pr\{X = 1\} = \lim_{N\to\infty}\left[\frac{N_{X=1}}{N}\right]$$

As the number of Bernoulli trials increases, the proportion of successful outcomes in the total number of trials tends to the actual probability of success. But this actually is the de Moivre's form of Central Limit Theorem that he published in 1733!

2.1.3 *Examples of SPC for Bernoulli Trials*

Attempts to establish a connection can be modeled as Bernoulli trials: They can either succeed or fail, and we are interested in the frequency (probability estimation) of successful attempts.

If we assign $p =$ Probability of success; $N =$ number of trials, then, given sufficient number of samples, CLT enables us to consider accessibility as a normally distributed metric with

$$\text{mean}_{\text{successful trials}}(n, p) = N * p \tag{2.1.3}$$

and

$$\text{variance}_{\text{successful trials}}(n, p) = N * p * (1 - p). \tag{2.1.4}$$

This will enable us to apply the SPC principles directly to attempts to establish connection. Two types of SPC charts, the p-chart and the np-chart (Fig. 2.1), which were originally proposed by Walter Shewhart and was named after him, are specifically designed for this kind of metric.

Collecting data for daily attempts to establish a connection will yield an np-chart like in Fig. 2.1.

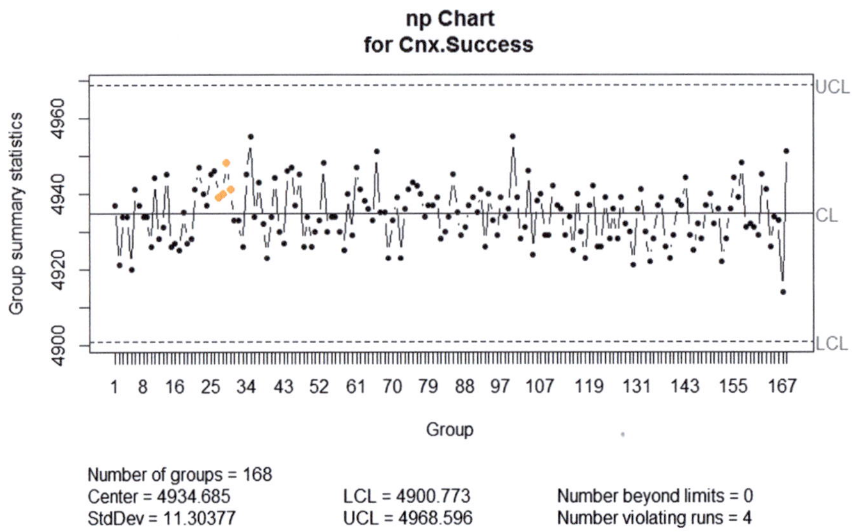

Fig. 2.1 Bernoulli trials: np-chart (simulated data)

Figure 2.1 shows the number of successfully established connections during each hour of one week. The control level (CL) is calculated using Eq. (2.1.3), while the Upper Control Limit and Lower Control Limit (UCL and LCL, respectively) are computed using the "3-sigma" rule:

$$\begin{aligned} \text{UCL} &= \mu + 3 * \sigma \\ \text{LCL} &= \mu - 3 * \sigma \end{aligned},$$

Here, $\mu = N * p$ = mean; $\sigma = \sqrt{N * p * (1 - p)}$ = standard deviation.

Figure 2.2 shows the number of successful ERABs established during each hour of one week. The control level (CL) is calculated using a form of Eq. (2.1.3) where both sides of the equation are divided by the number of samples (N), while the Upper Control Limit and Lower Control Limit (UCL and LCL, respectively) are computed using the "3-sigma" rule:

$$\begin{aligned} \text{UCL} &= \mu + 3 * \sigma, \\ \text{LCL} &= \mu - 3 * \sigma \end{aligned}$$

Here, $\mu = p$ = mean; $\sigma = \sqrt{\frac{p*(1-p)}{N}}$ = standard deviation.

The np- and p-charts enable the user to easily detect events when the data was out of control, e.g., the red data point at Group 87 (87th hour).

They also allow the analyst to find violations of the SPC rules of thumb, in particular Western Electric Rules (http://en.wikipedia.org/wiki/Western_Electric_rules) or Nelson Rules (http://en.wikipedia.org/wiki/Nelson_rules).

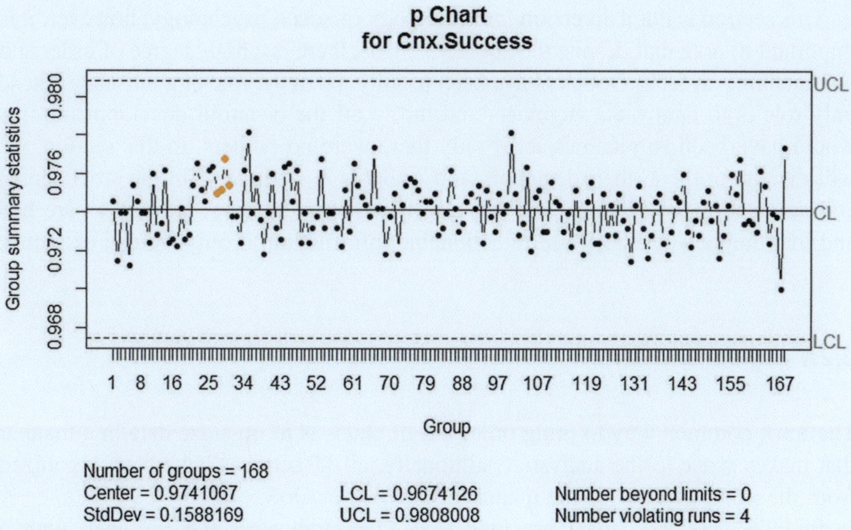

Fig. 2.2 Accessibility p-chart (simulated data)

Thus, the four yellow points in Figs. 2.1 and 2.2 correspond to violation of Nelson Rule 6 (or its equivalent Western Electric Rule 3): 4 out of 5 points fall outside the 1-standard-deviation band on the same side of the mean.

When the number of successful connections and the number of attempts is known for every hour, the R code to produce this graph is very simple:

```
##################################################################
## Plot the np-chart
##################################################################
if (! require ("qcc")){install.packages ("qcc")}
library (qcc)
npChart <- qcc (Cnx.Success, sizes = Attempts, type = "np")
pChart <- qcc (Cnx.Success, sizes = Attempts, type = "p")
```

2.2 Outliers

The "magic of LTE" involves a lot: A variety of users, sometimes with a number of different user elements per user, are trying to get access to the data. Uploads, downloads, VoIP communications, all requiring different connection times, different QoS, and different packet sizes, happen at the same time, ensuring that there is no limit to confusion and outliers.

This section is not a diversion into sociology or social psychology; however, it is important to note that despite the apparent chaos, there is a high degree of order and predictability in LTE. Outliers have been usually given the role of a nuisance whose only role is to annoy the engineers and throw off the beautiful developments that would have been so much easier if only there were no outliers. In this section, we will discuss outliers, their detection, and analysis in stationary (in the strict sense) and non-stationary systems. We will also try to show that not all outliers are bad and that they always carry useful actionable information, if only we can extract it.

2.2.1 *QoS Outliers*

The most common way to bring order out of chaos is to organize data in a manner that makes sense to the analyst. Traditionally, all IT services have been organized from the point of view of the quality of service, or QoS.

In LTE networks, there are nine QoS class indicators (QCIs), each with a specification on a packet delay budget and error loss rate (see Chap. 1 and Fig. 2.3).

The predefined packet delay budgets for each QoS class mean that an outlier may very well put a bearer into a lower-priority class, at least for the outlier. For example, a Priority 1 (QCI = 5, delay = 100 ms) connection may too easily fall into Priority 6 (QCI = 6, delay budget = 300 ms).

While packet error loss rate may be approximated as a normal distribution (see discussion in Sect. 2.1), packet delays are strongly right-skewed: Their distribution

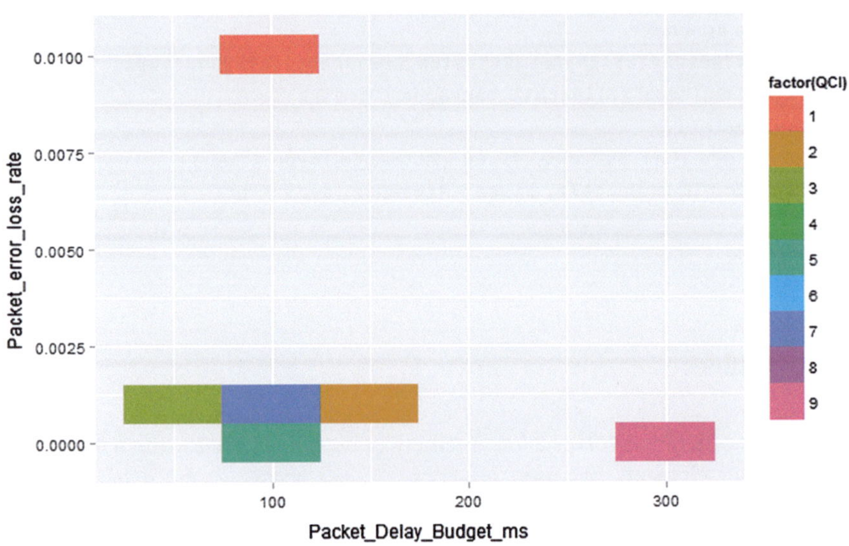

Fig. 2.3 LTE QoS class indicators (QCIs) defined by loss and delay specifications

has a long tail to the right. This phenomenon is often approximated by an exponential distribution. This makes the packet delay extremely sensitive to outliers.

Before we take a deeper dive into the topic, it is important to note that the concepts of outliers and of SLA (QoS) violations are, generally speaking, independent: Not all outliers are in violation of SLA and not all events of SLA violation are outliers.

2.2.2 *Outliers: What Are They?*

According to www.dictionary.com, an outlier in its statistical sense is

> … an observation that is well outside of the expected range of values in a study or experiment, and which is often discarded from the data set…

And here lies one of the most common temptations that we face when dealing with real-world data: The dictionary merely states it, but all too often we tend to discard outliers, even though more often than not, they are "in the eyes of the beholder."

Sometimes outliers do represent bad data collections, mangled data points, and simply unreliable data. However, when we know that the data is good, outliers become are important observations that can give us an advanced warning that the system's performance is beginning to degrade.

When the number of outliers for a KPI reaches an alarming level (again, "in the eyes of the beholder," but at least it can be quantified and parameterized), we want to continue monitoring the KPI specifically for the relative and absolute growth in outlier count and magnitude, to be able to predict when the performance will degrade if the system is left alone and to take appropriate corrective action.

Usually, the beholder—a subject-matter expert (SME)—knows what percentage of outliers in a sample the system can handle. Such knowledge is often guarded by the SME and eventually becomes "tribal knowledge" in the organization.

When dealing with thousands of objects providing data, the beholder's eyes become too strained and the beholder too bored analyzing each and every data set for outliers. In other words, an automatic outlier detection tool has to be implemented. Such tools are usually based on proven quantitative analysis (QA) techniques, which we shall consider here.

2.2.3 *Outlier Detection: The Basic Approach*

The SPC (see Sect. 2.1) represents outlier detection by applying the density-function method. As we discussed in Sect. 2.1, the mathematically simplest way to detect outliers is by using the mean (average) and standard deviation of the data. We then apply a multiplier to the standard deviation and get the confidence

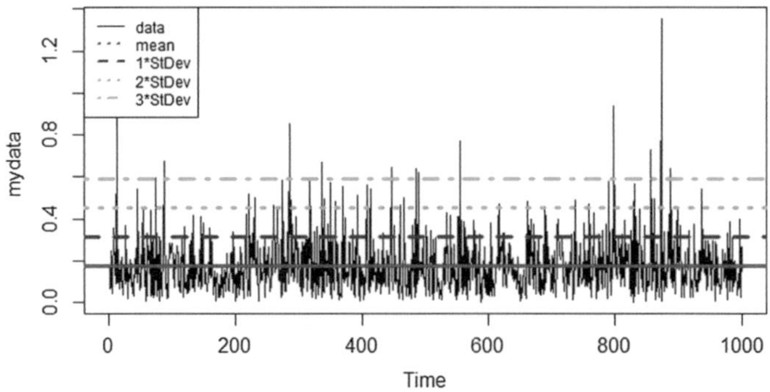

Fig. 2.4 A simulated right-skewed data set—representative of a typical latency (packet delay) observed on an LTE network—and its 0, 1, 2, and 3 standard deviations

interval and state that points outside such confidence interval (usually 3 standard deviations) are outliers. This is known as the "3-sigma rule."

Figure 2.4 shows a set of simulation data that we built to illustrate the concept. This view of the data allows us to identify outliers at different confidence levels:

$\pm 1 * \sigma$	68.2%
$\pm 2 * \sigma$	95%
$\pm 3 * \sigma$	99.7%

In other words, if a data point falls outside the $1 * \sigma$ range, we say with 68.2% confidence that this data point is an outlier. If another data point falls outside the $3 * \sigma$ range, that means that we are 99.7% certain that this data point is an outlier.

The code implementing these principles is here:

```
###############################################################
## Returns outlier indices in vector X corresponding to Z*StDev
###############################################################
GetZRangeOutlierIndices = function(X, Z = 3, Upper = TRUE, Lower = TRUE){
 # Upper = TRUE; Lower = TRUE ### for debugging
 Mean <- mean (mydata); SD <- sd (mydata)
 upper <- Mean + Z * SD
 lower <- Mean - Z * SD
 UpperOLindices <- which (Upper & X > upper)
 LowerOLindices <- which (Lower & X < lower)
 Outliers <- c(LowerOLindices, UpperOLindices)
 Outliers
}
##################################################################
## Generate 1000 samples of gamma-distributed "packet delay" data:
mydata <- rgamma (1000, shape = 1.7, rate = 10)
```

```
####################################################################
## Get outlier indices in mydata corresponding to 1, 2, and 3 StDev
####################################################################
zOLs.1 <- GetZRangeOutlierIndices(mydata, Z = 1)
zOLs.2 <- GetZRangeOutlierIndices(mydata, Z = 2)
zOLs.3 <- GetZRangeOutlierIndices(mydata, Z = 3)
```

We intentionally did not show the left side of the standard-deviation bands: The data is right-skewed, and for this particular analysis, we are not interested in low outliers.

Advantages of this approach are obvious:

1. It is mathematically simple.
2. It is fast.
3. Measures (mean and standard deviation) can be computed recursively, hence
   (a) Simple data storage, regardless of size.
   (b) Simple computations, regardless of size.
4. It makes sense statistically, giving answers in terms of confidence intervals.
5. It makes sense philosophically: Everybody knows what confidence is.

Disadvantages are fewer, but they can overwhelm the advantages:

1. Assumptions have to be made about the distribution.
2. The method will not work when these assumptions are wrong.

2.2.4 *Advanced Methods of Outlier Detection*

Section 2.6.2 discusses advanced methods of outlier detection and demonstrates an example of using information from in network performance analysis.

2.3 A Few Words About Queueing Systems

An illustration of a basic queueing system is shown in Fig. 2.5. The number of queueing circuits available is defined by the spectrum range and number of channels occupied at any given time (Bandwidth Efficiency).

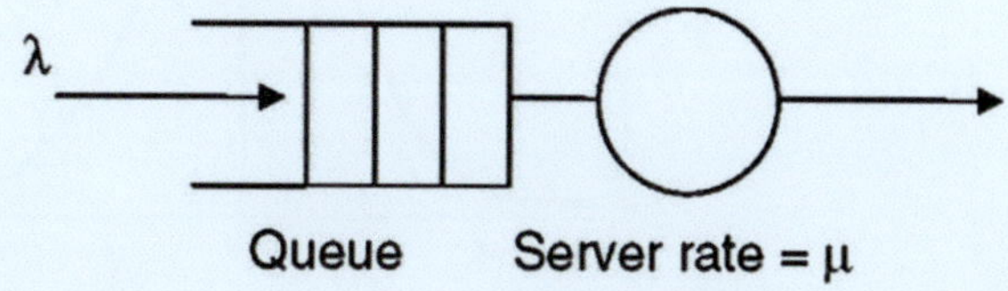

Fig. 2.5 A simple queueing system representing a EUTRAN channel

In this section, we will demonstrate why Central Limit Theorem is a hindrance, rather than help, in network performance analysis. When we are dealing with samples and sample averages, we will see them in data as normally distributed and apply statistical methods that will potentially lead us to wrong conclusions.

Indeed, for a queueing system with a single server (channel) and a buffering queue, the reason why we can describe the entire process easily via two variables, $\lambda = \text{rate}_{\text{arrivals}}$ and $\mu = \text{rate}_{\text{server}} = \frac{1}{T}$, is the assumption that they are not normally distributed (T = average service time).

Moreover, assumption of normal distribution would lead to unintended consequences. For example, a normally distributed service time (latency for a network) would have allowed negative latencies (Fig. 2.6).

In addition, even if the likelihood of negative latencies allowed by a specific normal distribution was negligibly small, the workload on the network element [measured as number of packets in flight and expressed as $W = \lambda * T = \lambda/\mu$] would come out skewed (Fig. 2.7).

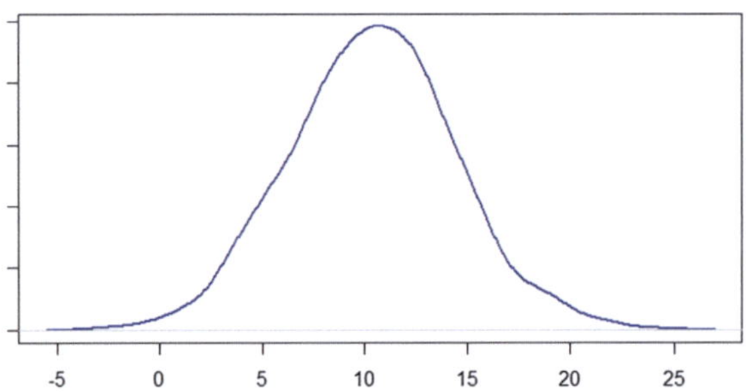

Fig. 2.6 A normal distribution allows negative latencies

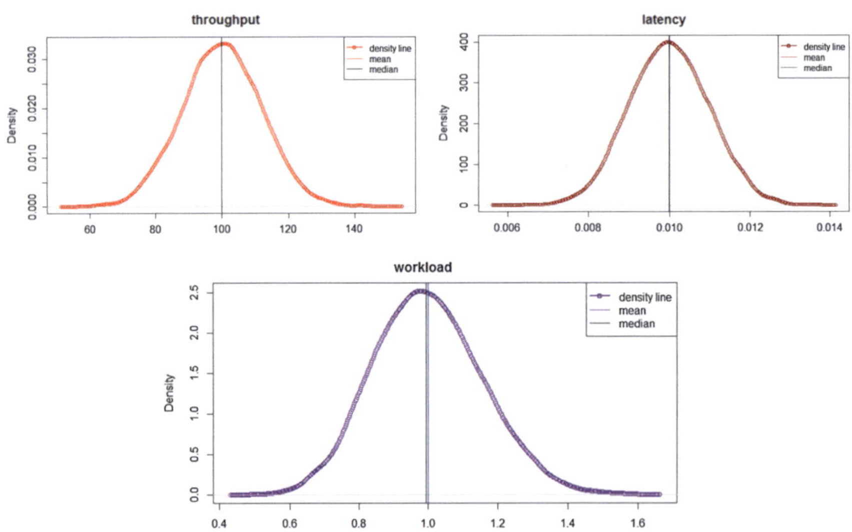

Fig. 2.7 An illustration of the skew in the product of two normal distributions: The gap between the *purple* (mean) and the *black* (median) *vertical lines* indicates that the distribution is skewed. For more information on skewness, see other literature

2.3.1 *"True" Process Distributions for LTE Network Components*

As Agner K. Erlang demonstrated in the 1920s, memoryless processes (where time intervals between adjacent events are independent of previous events) provide a very adequate representation of a queueing system. The only memoryless distribution known today is the exponential, and if we model network latencies using exponential distribution, we will have an adequate representation of the behavior of a network, including LTE network.

In order for event inter arrival times to be exponentially distributed, arrival process has to be Poisson.

Figure 2.8 illustrates these concepts in light of Little's Law.

This leads to workload distributions being non-normal as well.

This observation has two very important consequences: (1) if processes in an LTE link are memoryless, then Erlang models can be used to predict packet loss due to link overloads; (2) we cannot use any statistical methods created with the assumption of Gaussian (normal) distribution of underlying data.

Long story short, traffic metrics (throughput, latency, and workload) of a queueing system are not normally distributed.

2.3.2 *Little's Law: The "Big Three" of Queueing Dynamics*

A queueing-system dynamics can be fully characterized by three variables: (i) arrival rates, (ii) service rates, and (iii) workload. These are the "Big Three" of queueing dynamics.

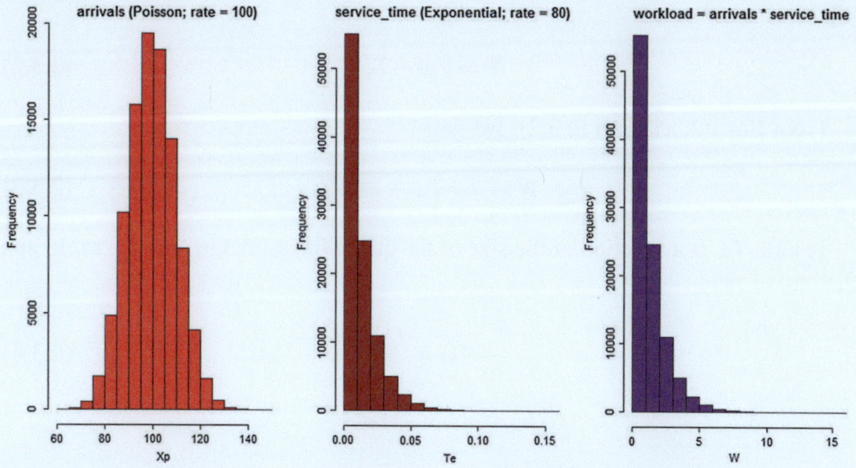

Fig. 2.8 The (throughput, latency, workload) tuple describing a queueing system

Packet Arrivals

Arrival rate is the characteristic of the demand on the queueing system.

Most commonly used models for sizing queuing systems today (Erlang models) rely on the assumption of certain prescribed distributions of arrival and service rates; in particular, that arrivals should be a Markov process, i.e., that job (packet) arrival rates should follow Poisson distribution.

For a long time, network engineering world has been rejecting the models derived for telephony on the grounds that Poisson distribution does not apply to network traffic, which is bursty and unpredictable. However, as our understanding of network, and in general IT traffic grew, so did our understanding that bursts could be modeled as Poisson arrivals of "trains" of packets, whose size (number of packets per "train") could be modeled as a Markov process. This opened an understanding of nested Markov arrival processes [IPPM2014], as well as made it possible to apply a quasi-Erlang model, approximating the nested Poisson distribution with a geometric distribution [FERR2014].

Modern methods of operations research (e.g., Monte Carlo analysis) open new opportunities for creating network analytical systems for LTE where traffic mixes with a variety of arrival processes can be analyzed from the perspective of the probability of delay and/or probability of blocking.

Service Process

Little's Law (Eq. 2.3.1) is a universal law of nature. It ties together the concurrency (number of units of work, e.g. jobs, queries, and packets) that are in the system at any given moment in time) with arrival rate and waiting time:

$$W = \frac{\lambda}{\mu} = \lambda * T \tag{2.3.1}$$

In turn, waiting time can be represented as the sum of the time in and the time in service:

$$T = T_{\mathrm{W}} + T_{\mathrm{S}}, \tag{2.3.2}$$

From Eqs. (2.3.1) and (2.3.2), we get:

$$W = \lambda * T_{\mathrm{W}} + \lambda * T_{\mathrm{S}} \tag{2.3.3}$$

In turn, T_{W} is a function of the size of the queue (measured in units of work) and the arrival rate:

$$T_{\mathrm{W}} = \frac{Q}{\lambda} \tag{2.3.4}$$

T_S is the inverse of service rate:

$$T_S = \frac{1}{\mu} \tag{2.3.5}$$

One very important property of service time for a network is that it depends on the packet size. In other words, while the bit processing rate is a constant for any network element, the number of packets that it can turn out varies:

$$\mu_P\,[\text{packets per second}] = 8 * \frac{\text{BW [bits per second]}}{S\,[\text{bytes per packet}]} \tag{2.3.6}$$

The bandwidth of a network element is a critical parameter defining the packet service time!

Putting it All Together

Putting (2.3.3)–(2.3.6) together, we have an expression for the number of concurrent packets in the network element:

$$W\,[\text{packets}] = \lambda\,[\text{packets per second}] * \left\{ T_W + \frac{S\,[\text{bytes per packet}]}{8 * \text{BW [bits per second]}} \right\} \tag{2.3.7}$$

This can be translated into:

$$W[\text{bits}] = \frac{8 * \lambda\,[\text{bits per second}]}{S\,[\text{bytes per packet}]} * \left\{ T_W + \frac{S\,[\text{bytes per packet}]}{8 * \text{BW [bits per second]}} \right\} \tag{2.3.8}$$

Generally speaking, the packet size, S [bytes per packet], is not the same for arriving packets as it is for the packets already in flight (in service). Simple algebra leads to:

$$\forall i \in \{\text{incoming packet types}\}:$$

$$W_i^S\,[\text{bits}] = \frac{8 * \lambda_i\,[\text{packets per second}]}{S_i^{\text{in}}\,[\text{bytes per packet}]} * \left\{ T_W + \sum_{f \in \{\text{packet types in flight}\}} \frac{S_f^{\text{out}}\,[\text{bytes per packet}]}{8 * \text{BW [bits per second]}} \right\} \tag{2.3.9}$$

In Eq. (2.3.9),

S_i^{in}	size of packet type i
W_i^S	number of packets in flight of type i

λ_i [packets per second] arrival rate of packets of type i
S_f^{out} size of packets in flight of type f
BW bandwidth of the network element

And total concurrency (number of bits in flight) will be:

$$W\ [\text{bits}] = \sum_{i \in \{\text{incoming packet types}\}} W_i^S\ [\text{bits}] \tag{2.3.10}$$

Equations (2.3.9) and (2.3.10) are very important: They allow us to compute the packet delay. Indeed, simple algebra yields:

$$\begin{aligned} T_{\text{W}} = & \sum_{i \in \{\text{incoming packet types}\}} \frac{W_i^S\ [\text{bits}] * S_i^{\text{in}}\ [\text{bytes per packet}]}{8 * \lambda_i\ [\text{packets per second}]} \\ & - \sum_{f \in \{\text{packet types in flight}\}} \frac{S_f^{\text{out}}\ [\text{bytes per packet}]}{8 * \text{BW}\ [\text{bits per second}]} \end{aligned} \tag{2.3.11}$$

Equation (2.3.11) is the general form of the waiting (queueing) time, which is the main component of packet delay. If we tie it back with Little's Law, it becomes mathematically obvious how to compute the buffer for packet delay.

2.3.3 *System Performance Laws*

Some LTE behavior patterns are unique to LTE. Others follow well-established laws of IT systems, many of them derived from queueing-system considerations.

2.3.3.1 Amdahl's Law

In 1967, Gene Amdahl formulated the argument against putting too much trust into parallel computing. According to this argument, which later became known as Amdahl's Law, there is a limit to the speedup that a computing system can obtain as a result of massive parallelization. This limit is controlled by the relationship between the number of parallel processors and the amount of serial processing in the program as percentage of total processing:

$$\text{Speedup} = \frac{1}{S + \frac{1}{N} * (1 - S)} \tag{2.3.12}$$

In Eq. (2.3.12), The S and the N are the fraction of processing that is strictly serial and the number of parallel servers, respectively.

2.3.3.2 Universal Scalability Law

In 1993, Neil Gunther formulated the Universal Scalability Law (aka USL), which took Amdahl's Law one step further and generalized it to the case of processors communicating within each other. For more details, we highly recommend [GUNT2007]. Gunther formulated the relationship between the system throughput and the number of concurrent users in the system:

$$C = \frac{N}{1 + \sigma * (N - 1) + \kappa * N * (N - 1)} \tag{2.3.13}$$

In (2.3.13), the C is the system throughput (capacity); σ and the κ are parameters describing contention for resource and incoherency (lack of communication) during the job execution, respectively.

Figure 2.9 shows how the USL curve changes with different USL parameters.

Depending on the system's performance parameters (contention and incoherency), the USL curve will grow more or less steep, level off at some number of concurrent users, or peak and start going down having reached the maximum number of users it can handle.

In the recent years, there have been a number of improvements to USL and its "cousin," the Superserial Scalability Law (SSL); in particular, [CHDY2014] suggested a third parameter, β, to account for queueing:

$$C = \frac{N}{1 + \sigma * (N - 1) + \kappa * N^{\beta} * (N - 1)} \tag{2.3.13a}$$

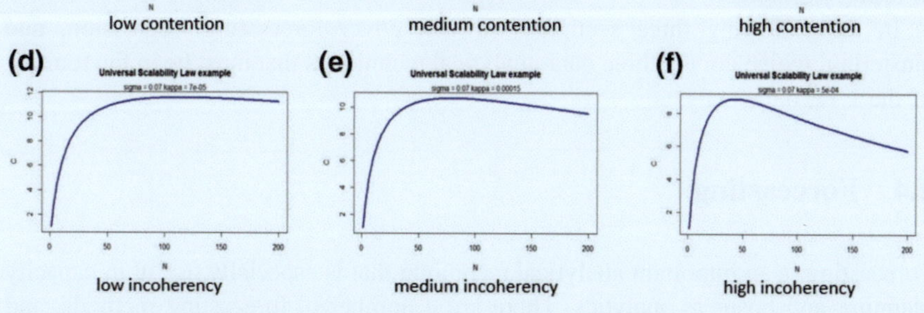

Fig. 2.9 Universal scalability law examples

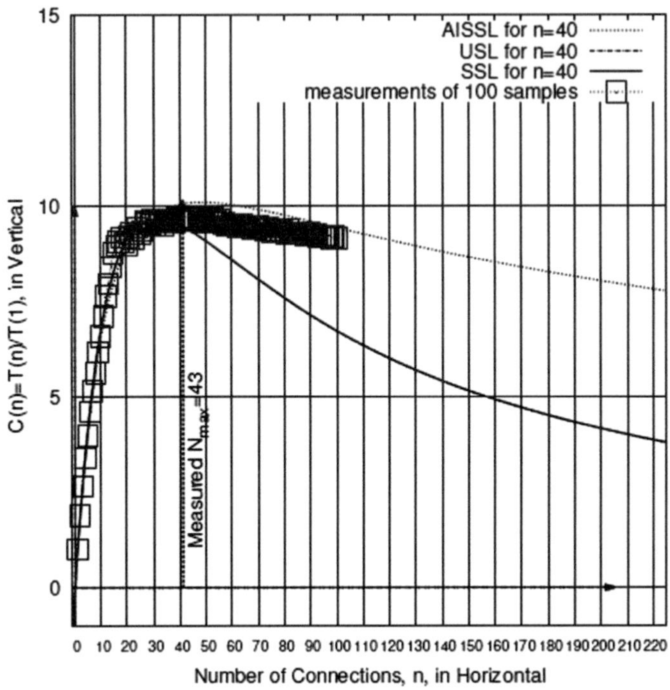

Fig. 2.10 Different versions of USL and SSL and real-world data (reproduced from [CHDY2014] with author's permission)

The Choudhury's AISSL (Asymptotically Improved SSL) form better fits real-world data by smoothing the sharp drop predicted by USL in Fig. 2.9f (Fig. 2.10).

2.3.4 Conclusion

In this section, we have covered queueing process distributions, Little's Law, and scalability laws (Amdahl's and the USL family) and described the packet delay due to retransmissions.

In the following three sections, we will cover forecasting, regression, and clustering, which are the three data-analytical techniques that must be in the toolbox of the LTE data analyst.

2.4 Forecasting

Forecasting is an important analytical technique that is especially useful in capacity planning and business analytics. There are a number of forecasting methods, and each of them deserves a separate book in its own right. In this book, we will only cover forecasting techniques formulated for time-series analysis (TSA). However, in Sect. 2.5, we will discuss regression and its use in forecasting.

2.4.1 Time Series: Definition and Assumptions

A time series is a set of system metric measurements sampled in constant time intervals and arranged in the order of the time stamps. In TSA, assumptions are made that there are no missed data and that each metric's time series can be viewed as independent of other metrics.

In TSA forecasting, the additional assumption is made that the system's behavior in the past will continue into the future. While the first to assumptions can be justified or enforced, this last one is not as trivial.

Consider the system in Fig. 2.11. It can be any MISO system, e.g., two big groups of users as the input and SINR as the output.

Note that while Fig. 2.11 shows a dependency between the inputs and the output of the system, and the two inputs most likely are interacting (e.g., multiplexing), as long as the processes described by the system are stationary, assumption of independence will hold true for all 3 signals. This can be confirmed by running a number of statistical tests, from correlation tests to ANOVA. For more details, see earlier sections of this chapter, as well as other literature. As long as it is the case, we can apply the forecasting techniques that we are introducing in this section to predict future behaviors of $X1$, $X2$, and Y.

However, if a user control is introduced (Fig. 2.12), the situation will change.

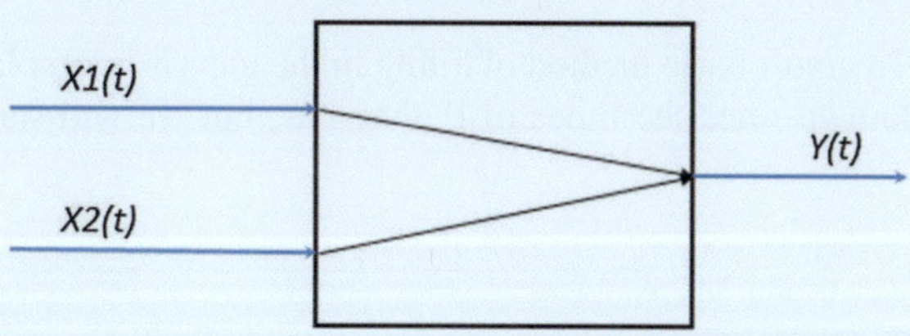

Fig. 2.11 A system with 2 incoming and one outgoing signals

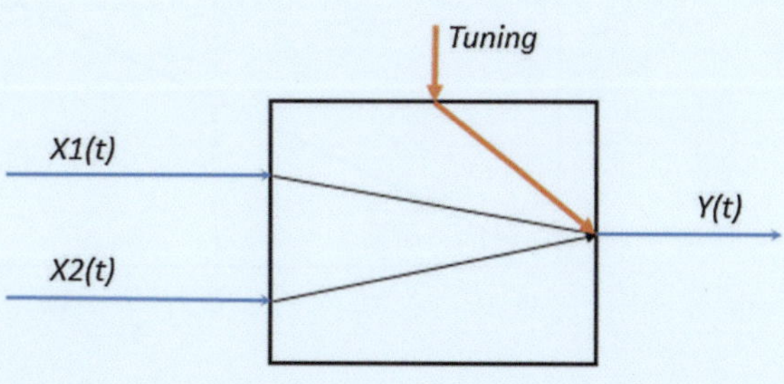

Fig. 2.12 The same system with 2 incoming and one outgoing signals

Now, we have introduced a tuning parameter, e.g., the azimuth of the tower's antenna. Now, the SINR has changed, and the predictions made for *Y*(*t*) based on previous behavior will not be adequate anymore.

2.4.2 *Filling in the Gaps in Data*

2.4.2.1 Why?

All TSA techniques are based on having constant time intervals between sequential measurements and the data being ordered by time stamps. On the most basic level, a gap in data means that we have no knowledge about what was happening to the system during this time interval.

2.4.2.2 Methods: Extension

The most primitive method for filling in the gaps in data is extension of the last point's value that preceded the gap, aka LOCF—Last Observation Carried Forward. It is very effective and "unassuming" when it comes to small gaps that need to be filled. The downside is that the value changes in a step function (see the step at X = 30 in Fig. 2.13), which sometimes may mean an actual level shift and sometimes a gap filled by extension.

2.4.2.3 Methods: Interpolation

The most basic method of filling in the gaps is interpolation. The technique has been known since the times of Pythagoras, but we will start with it.

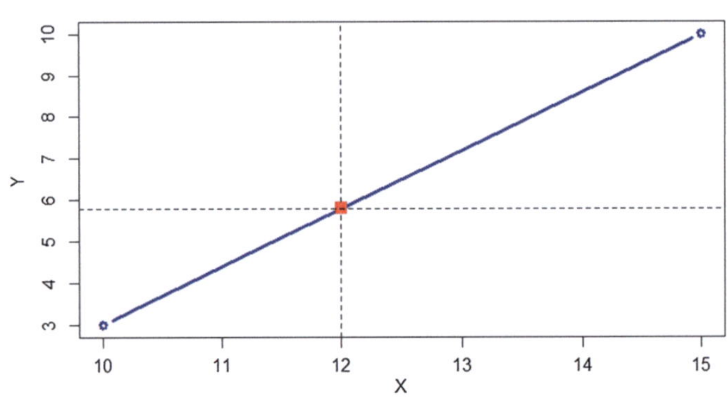

Fig. 2.13 Interpolation

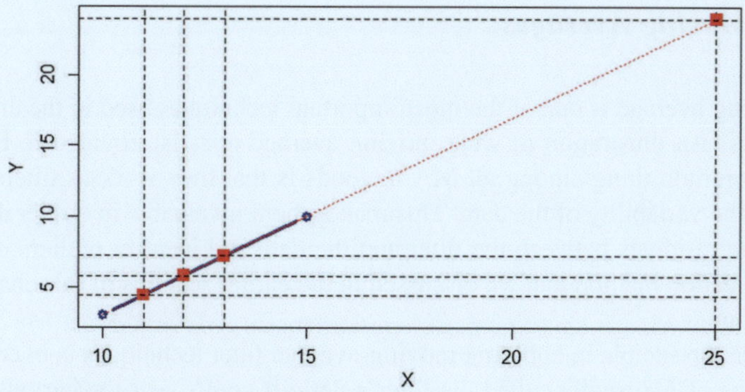

Fig. 2.14 Multi-point inter- and extrapolation

If Y is the value of the dependent variable that we need to find given X—the value of the independent variable, and we are given $(X_1, Y_1), (X_2, Y_2)$ pairs, then:

$$Y = Y_1 + (X - X_1) * \frac{Y_2 - Y_1}{X_2 - X_1} \tag{2.4.1}$$

Naturally, the formula (2.4.1) works just as well for multiple values of X and even for points outside the initial $[X_1, X_2]$ range in which case it is known as extrapolation (Fig. 2.14).

If we are dealing with a time series, interpolation gives us a simple tool to deal with missing data: If X is the point ID (or time stamp), and Y is the value we have measured on the system, and some X values do not have a corresponding Y, we can use interpolation to estimate the missing values (Fig. 2.15).

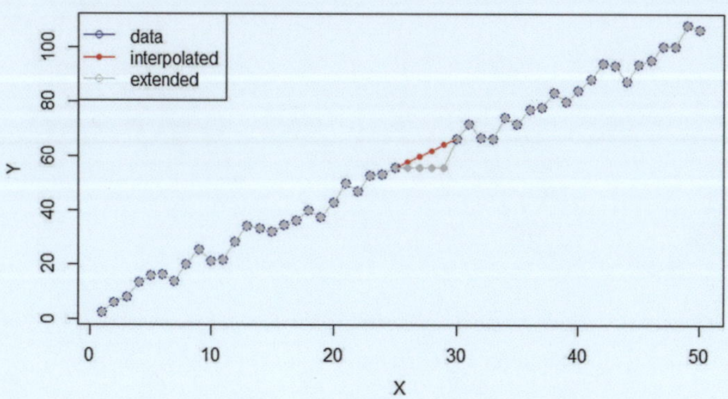

Fig. 2.15 Time series with extension and interpolation of missing data

2.4.3 Moving Averages

The moving average is one of the most important techniques used in the analysis of time series. An illustration of what moving average does is provided in Fig. 2.16.

One common thing among all MA methods is that they work as filters, greatly reducing the variability of the data. This makes them invaluable in outlier detection: look at the residuals between the filter and the data and identify outliers using any outlier detection method that we discussed in the earlier sections of this chapter, and we are done.

The main principle in applying moving-average filter techniques is in computing the average of *N* (usually called the filter's period) points near the current point. It may be *N* points behind, *N* points centered on the current point, or *N* points in front of it. In TSA forecasting, it is usually *N* points behind.

To improve the performance of the MA filter, each point is endowed with a weight, with the idea that the points far in the history should have relatively little say in the behavior of the filtered data, whereas more recent data points should carry a lot more weight.

The techniques produce different curves. Closest to the data line (*Y* in Fig. 2.16) are typically SMA (simple moving average) and WMA (weighted moving average). The SMA in time-series analysis is also known as "random walk," because of its sensitivity to the stochastic component of the data. The most sensitive to trending and periodic patterns in data is the EWMA, aka EMA (exponentially weighted moving average), aka exponential smoothing. The LWMA (linearly weighted moving average) is in between these two.

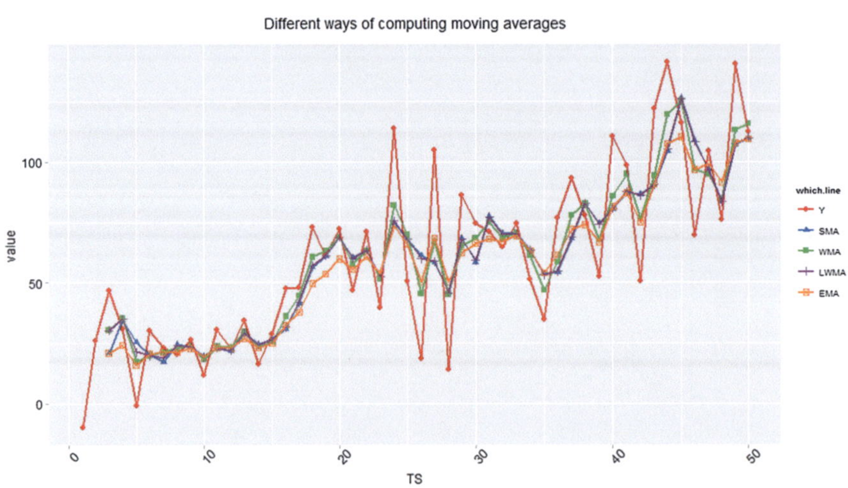

Fig. 2.16 An illustration of different ways of computing moving average

2.4.4 EWMA

The word "exponential" in "exponential smoothing" causes a lot of confusion outside the data-analytical community. It has nothing to do with the trend being exponential, linear, or even flat. The EWMA model computes the next value based on weighted previous data points in a manner that can be recursively written as shown in Eq. (2.4.2):

$$\begin{aligned} y_0 &= x_0 \\ y_t &= \alpha * x_t + (1-\alpha) * y_{t-1}, \quad \forall t > 1 \end{aligned} \tag{2.4.2}$$

In Eq. (2.4.2),

x the time series;
y the fitted model;
$0<\alpha<1$.

Equation (2.4.2) can be rewritten as:

$$\begin{aligned} y_t = \alpha * &\left[x_t + (1-\alpha) * x_{t-1} + (1-\alpha)^2 * x_{t-2} + \cdots + (1-\alpha)^{t-1} * x_1\right] \\ &+ (1-\alpha)^t * x_0, \quad \forall t > 1 \end{aligned}$$

Or in a more compact form:

$$y_t = \alpha * \sum_{i=0}^{t} (1-\alpha)^{t-i} * x_i + (1-\alpha)^t * x_0, \quad \forall t > 1 \tag{2.4.3}$$

It is because of the exponential form of the $(1-\alpha)$ weights that the method is called exponentially weighted moving average.

2.4.5 EWMA Forecasting

The only limitation on the t in Eq. (2.4.3) is that it should be greater than 1. This means that we can extend the series using EWMA forecasting to any point in the future.

An illustration is shown in Fig. 2.17.

Prediction Interval

The two nested gray envelopes around the blue line in Figs. 2.17 and 2.18 are measures of forecast precision. They outline the zones where 60% (dark gray) and 90% (light gray) of the data will likely fall for each time stamp in the forecast horizon.

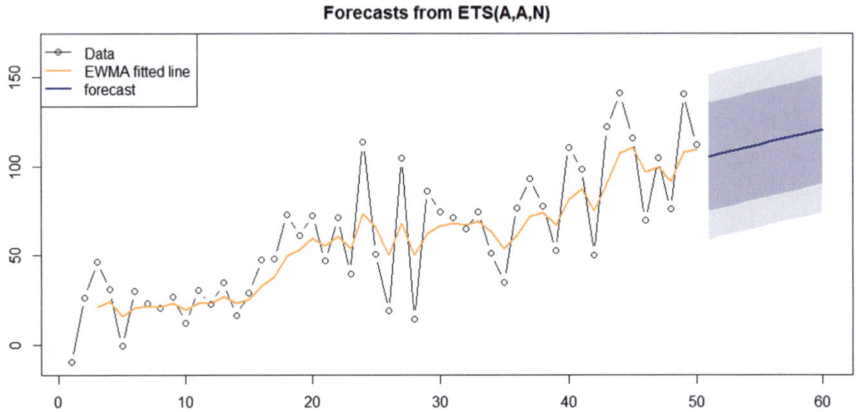

Fig. 2.17 EWMA forecast of the time series shown in Fig. 2.16

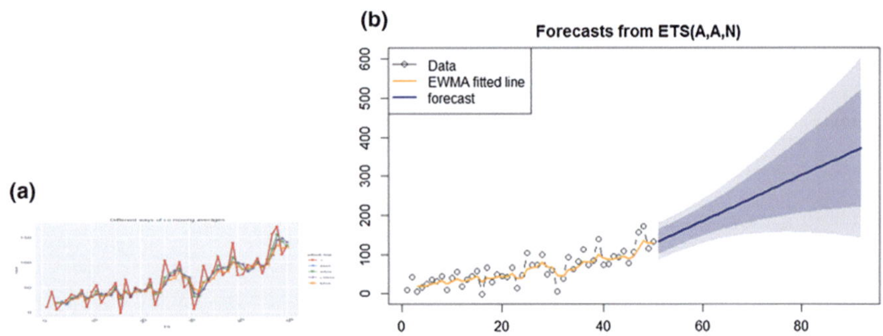

Fig. 2.18 A time series with a runaway prediction interval: **a** the four moving-average filters applied to the time series, **b** the EWMA forecast

In Sect. 2.1, when we discussed SPC, we talked about confidence intervals. For forecasts, the term "confidence interval" is replaced by "prediction interval." It is computed based on the model residuals (for a more in-depth discussion of model residuals, reader is referred to Sect. 2.5).

The "any point in the future" should be taken with a grain of salt: Forecasting a time series into a horizon bigger than the size of the historical data set is ill-advised. Typical forecast horizons are usually between half and 90% of the length of the historical data.

Consider the time series in Fig. 2.19.

Here, the forecast prediction interval is growing as we are moving farther away from the last historical data point, and by the time the forecast horizon (42 points) is reached, the upper 95th percentile of the prediction interval is already in the low 600 s, while the lower 95th percentile is still in the low 140 and even below the value of the time series at the last historical data point.

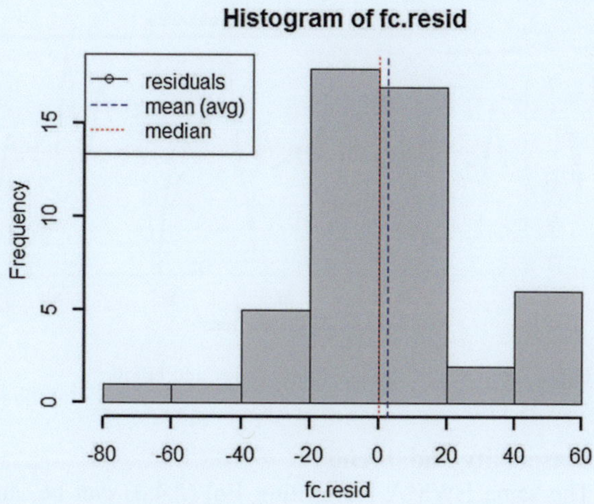

Fig. 2.19 Histogram of the EWMA forecast residuals

About Point Forecasts and Model Residuals

A point forecast is a prediction of the value being forecasted that does not have a prediction interval associated with it: Each time stamp of the forecast horizon is a single point. In Figs. 2.17 and 2.18, it is represented by the blue line.

It is important to remember that regardless of what shape the forecasted line takes, it is not meaningful to talk about a point forecast: Data volatility necessitates ranges of uncertainty around the centerline of the forecast.

Forecast residuals are computed as

$$\text{resid} = \text{data} - \text{forecast}$$

Their values are an important measure of how well the model fit the data. For the forecast shown in Fig. 2.18, the histogram of the residuals is shown in Fig. 2.19.

We see from the histogram that (i) the distribution of the residuals is not normal (it is compressed on both sides); (ii) there is a big bulge in the distribution where the tail usually is; and (iii) the forecasting model was anchored to the median (the red vertical line is zero) and not to the mean of the data.

Boxplot (see details in Sect. 2.2) reveals that we have outliers, and plotting the residuals as a time series and drawing the outlier boundary lines ("whiskers") will help us identify them (see details in Sect. 2.1). These concepts are illustrated in Fig. 2.20.

Measures of Forecast Model Quality

A number of measures of model quality have been developed over the years that EWMA forecasting has been in operation. Any forecasting tool can be tuned to optimize any of these statistics, reducing the error margins. The forecast we analyzed has been anchored to the median, but it could have been anchored to the mean of the historical data.

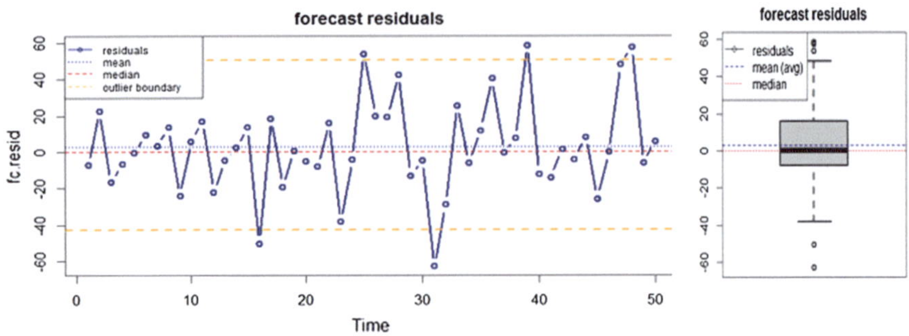

Fig. 2.20 Forecast residuals time series and boxplot

Seasonality and Trend

The same EWMA forecasting Eq. (2.4.3) can be, and have been, generalized to forecast trend and seasonality. The idea is the same: adding exponentially weighted values of the historical time series to calculate the trend and the seasonality at each point of the horizon (the so-called additive trend and additive seasonality). Moreover, they can be, and have been, modified by replacing "+" with a "*" to implement multiplicative seasonality and trend in EWMA forecasting.

Seasonality in time-series analysis is not necessarily tied into the four seasons. Any periodic pattern in a time series is called seasonality. It can be annual, semi-annual, quarterly, monthly, weekly, diurnal, and even hourly, but these are absolutely not the only periodic patterns that can be observed in the data. Seasonality in time-series data reveals a great deal of critical information about the dynamics of the processes that the data was collected from and can help control these processes.

If the magnitude of the seasonal changes does not increase or decrease in time, such seasonality is called additive, and Eq. (2.4.3) can be used with it "as is." When the magnitude of seasonal changes increases or decreases consistently, then it is called multiplicative seasonality, and the modified form of Eq. (2.4.3) needs to be applied to such data.

Seasonality and Trend Detection

It is not the purpose of this book to focus on seasonality and trend detection methods, but cluster-based seasonality analysis [GILG2010(1)], Fourier analysis, autocorrelation, and other techniques have proven themselves to be very useful in the field, leading to great improvements in the quality of the forecasts. This area, as well as its cousin change-point detection, is very rapidly growing areas in statistics, and one should always keep an eye on new developments there.

2.4.6 ARIMA Forecasting

"ARIMA" stands for "AutoRegressive Integrated Moving Average." It too is a moving-average technique, where the seasonality and the trend are detected automatically via time-series analysis. It is also known in time-series analysis as the Box-Jenkins forecasting method.

"AR"
Instead of applying an exponentially changing weight to the data points, ARIMA seeks to build a linear regression of each point's value to all of its predecessors and uses the coefficient of such model as the weights applied to each data point:

$$Y_t = c + \sum_{i=1}^{p} \varphi_i * X_{t-i}$$

The p is the number of degrees of freedom for the "AR" element of the ARIMA.

"I"
One problem with autoregressive models is they require the time series to be stationary (mean and variance to be "close enough" to constant). In the real world, it is not the case. One way to deal with it is to "difference" the time series:

$$X'_{\mathrm{t}} = X_{i+1} - X_i$$

The number of times it takes to make a time series stationary is usually designated as d and is the parameter of the "I" part of ARIMA.

"MA"
The "MA" (moving average) part works the same way as the already described SMA. Its parameter is usually designated as q and implies the width of the moving-average window.

Putting It All Together for ARIMA
When we apply ARIMA to the time series we have been examining in this section (Fig. 2.16), we have to be careful as to what values for (p, d, q) parameters we use. The parameters for most forecasting problems stay within the range of [0…2] (Fig. 2.21).

Let us explore the variety of ARIMA forecasts with different parameters.

2.4.7 Selection of Forecasting Model

The best overview of methods used in selecting a forecasting model is in [MAKR1998].

The model is usually selected based on how well it fits the data. The most common general strategy is similar to the approach used in the well-known, yet

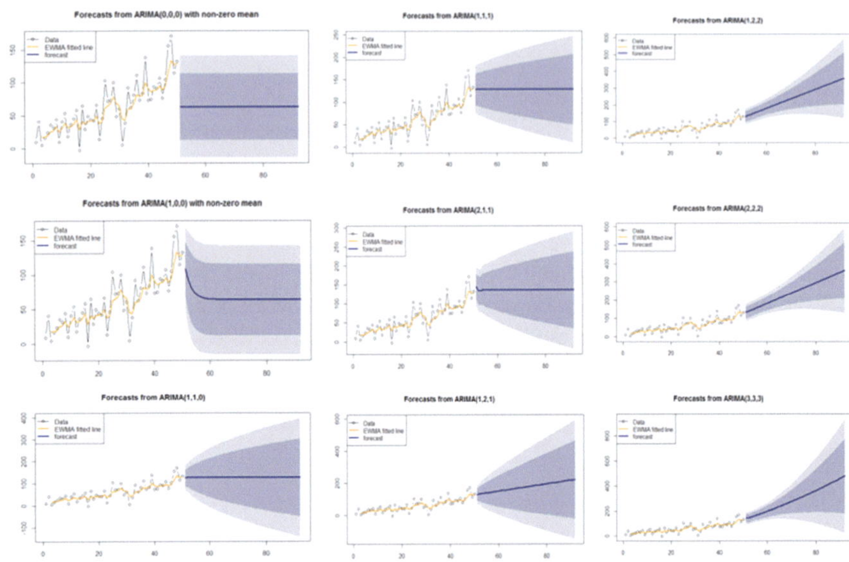

Fig. 2.21 ARIMA forecasts with different parameters

little understood, R^2 metric: compare model prediction with the data in every point of historical data and formulate the metric based on this comparison. Other approaches, such as out-of-sample testing, are used as well.

One of the problems is the variety of model quality evaluation methods, and finding ways to deal with it is one of the frontiers in TSA.

A weighted-sum synthetic approach to formulating the TSA model score has been successfully implemented, e.g., by [GILG2010(2)].

Regardless of what is chosen to be the model quality metric, two basic approaches are now in use in the predictive-analytics world: (i) select the model that fits best and (ii) use a so-called ensemble method. The latter relies on a Delphi approach, where several models are applied, analyzed, and scored, and then, their predictions are averaged with weights assigned based on their scores.

2.5 Regression

… all models are wrong, but some are useful.

—G.E.P. Box

Regression is another important tool in the network analyst's toolbox. It is the link between exploratory and predictive data analytics. As part of the exploratory data analysis (EDA), we always want to see whether we can find any correlations among the variables.

For example, if we see diurnal variations in throughput, we have an understanding that it is most likely driven by the number of user elements and their demand on the network.

This leads to several interesting observations: (i) We can forecast throughput using TSA techniques; (ii) we can also forecast throughput using business metrics (UE count and per-user throughput); (iii) if indeed there is a correlation between these two business metrics, it becomes a function of two variables: user count and per-user throughput.

In turn, total throughput is a nonlinear function of the number of users (see Sect. 2.3, Queueing Theory), which makes a seemingly simple task of predicting total throughput on a network a daunting one, due to all the interdependencies.

2.5.1 A Few Words on Terminology

The term "linear" is overloaded when it comes to regression. Indeed, there is an "official" mathematical definition of the term: http://en.wikipedia.org/wiki/Linearity, but this definition implies that an equation of the form

$$y = a * x + b, \tag{2.5.1}$$

(a, b = parameters; x, y = variables) is already nonlinear: the Homogeneity of Degree 1 is not true, unless $b = 0$:

$$m * (\alpha * x) + b = \alpha * m * x + b \neq \alpha * (m * x + b) = \alpha * m * x + \alpha * b$$

Nevertheless, when describing a regression line of the form (2.5.1), we call it linear, the reason being that its right-hand side is a linear polynome *in parameters*. By this definition, a quadratic, cubic, and any higher-degree polynomial function of x of the form $y = a_0 + a_1 * x + a_2 * x^2 + \cdots + a_n * x^n$ will be linear for all intents and purposes of regression: Each parameter $a_0, a_1, \ldots, a_n$ is entered in this polynomial as linear elements.

2.5.2 Linearizable Relationships

The "Big 6" of most commonly used linearizable relationships are as follows:

1. Straight-line relationship: $y = a_0 + a_1 * x$
2. Quadratic relationship: $y = a_0 + a_1 * x + a_2 * x^2$
3. Exponential relationship: $y = a_0 * e^{a_1 * x}$
4. Logarithmic relationship: $y = a_0 + a_1 * \ln(x)$

5. Power relationship: $y = a_0 * x^{a_1}$
6. Hyperbolic relationship: $y = \frac{a_0}{a_1 + a_2 * x}$

(Polynomials of higher degree than quadratic are very rarely used in regression analysis: Occam's razor demands that models should be as simple as possible, and cubic parabola and above rarely meet that requirement.)

We call these 6 linearizable because by simple substitution, we can convert them to linear:

1. Straight-line relationship: $y = a_0 + a_1 * x$:
   (a) no need to change anything
2. Quadratic relationship: $y = a_0 + a_1 * x + a_2 * x^2$:
   (a) Substitute $x_1 = x; x_2 = x^2$
   (b) $y = a_0 + a_1 * x_1 + a_2 * x_2$
3. Exponential relationship: $y = a_0 * e^{a_1 * x}$:
   (a) Take In of both sides
   (b) Substitute $y_1 = \ln(y)$
   (c) $y_1 = \ln(a_0 * e^{a_1 * x}) = \ln(a_0) + \ln(e^{a_1 * x}) = b_0 + a_1 * x$
      i. $b_0 = \ln(a_0)$
4. Logarithmic relationship: $y = a_0 + a_1 * \ln(x)$:
   (a) Take exp of both sides
   (b) Substitute $y_1 = e^y$
   (c) $y_1 = e^{a_0 + a_1 * \ln(x)} = a_0 + a_1 * x$
5. Power relationship: $y = a_0 * x^{a_1}$:
   (a) Take In of both sides
   (b) Substitute $y_1 = \ln(y); x_1 = \ln(x)$
   (c) $\ln(y) = \ln(a_0) + a1 * \ln(x) \Leftrightarrow y_1 = b_0 + a_1 * x_1$
      i. $b_0 = \ln(a_0)$
6. Hyperbolic relationship: $y = \frac{a_0}{a_1 + a_2 * x}$
   (a) Divide numerator and denominator by a_0: $y = \frac{1}{b_1 + b_2 * x}$:
   (b) Take the inverse of both sides: $\frac{1}{y} = b_1 + b_2 * x$
   (c) Substitute $y_1 = \frac{1}{y}$
   (d) Finally, $y_1 = b_1 + b_2 * x$

2.5.3 The Main Idea Behind Regression

In Fig. 2.22, we see a scatter plot around the line $y = 5 + 1.5 * x$.

The scatter (jitter/noise) around the solid black line is the stochastic (i.e., random) component of the data. In the real world, we usually know the nature of the relationship between the X and the Y, but do not know the equation of the solid black line; so we try to approximate (infer) it from the data. We want to draw a line that will minimize the error of such approximation.

Several measures of error have been proposed, the most prominent one, and the oldest one of all, being the sum of squared errors. We draw a line

$$y' = a_0 + a_1 * x: \quad (a_0, a_1) = \arg\min\left[\sum (y - y')^2\right] \qquad (2.5.2)$$

In Eq. (2.5.2), y' = fitted data; y = measured data.

This form of regression has got the name of *Least-Squares regression*; sometimes, it is called *Ordinary Least-Squares (OLS) Regression* and is what is commonly referred to as *linear regression.*

By applying the transformations outlined in Sect. 2.5.2, and expanding it to a multivariate form (hint: the quadratic relationship has been converted to a bivariate form), we can model a wide class of relationships between the X and the Y.

The X and the Y are usually referred to as independent and dependent variables, respectively. Often in multivariate problems, the Xs are called "covariates," and in machine-learning applications, the Xs are usually referred to as "features."

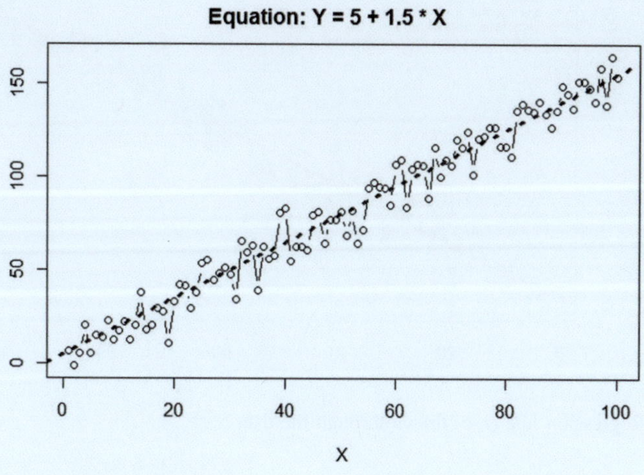

Fig. 2.22 Main idea behind regression

2.5.4 *Solving Eq. (2.5.2)*

To solve Eq. (2.5.2) is to fit a regression line into the data. Mathematically, it is done by solving a linear matrix equation which is outside the scope of this book. Interested readers are referred to any text in basic statistics. We are focused on the applied aspects of regression.

For the scenario shown in Fig. 2.22, the best fit we could get with OLS regression is shown in Fig. 2.23.

We see that the equation is off.

The intercept for the best-fitted model is $a_0 = 2.69$, while the coefficient is $a_1 = 1.53$. To verify that this error is induced by the noise, we can fit a regression line into the data produced by simply generating the data using the equation $Y = 5 + 1.5 * X$, and such regression's parameters will be exactly $a_0 = 5.00$ and $a_1 = 1.50$.

This error is what forces the fitted line (the red dashed line in Fig. 2.23) to be off the black solid line there.

Caution

Do not try to predict the line in the reverse order: A regression built for $Y(X)$ will not necessarily work to predict $X(Y)$, even with the algebra done correctly. The reason for that is in the way OLS parameters are computed and is a consequence of the laws of matrix multiplication.

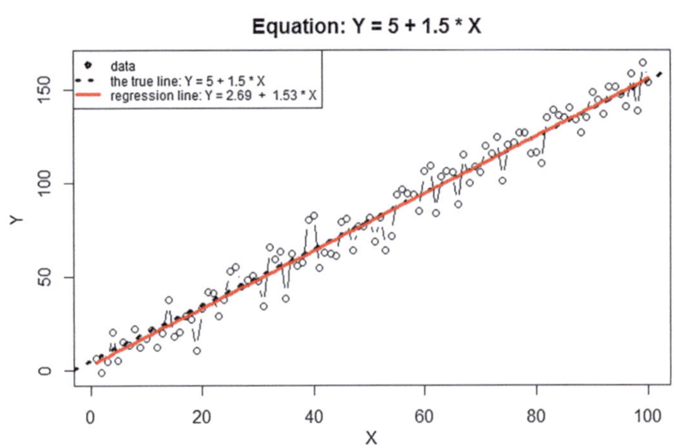

Fig. 2.23 Regression line (*red*) drawn through the data

2.5.5 Goodness of Fit

R-Squared

The most familiar to all goodness-of-fit estimator is the R^2. Created as a nonnegative version of ρ-Pearson's correlation factor, R^2 represents the proportion of the data variance that has been explained by the model. For more mathematical details, interested readers are referred to, e.g., [MILT2002]. Being a square of Pearson's correlation factor, it has some properties that make it preferred over the ρ. Namely, Pearson's ρ is formulated so as to indicate the directionality of the correlation ($\rho > 0$ if Y increases with X, and $\rho < 0$ if Y decreases as X grows). The R^2 eliminates this issue, making it better suited as a metric of model quality.
One of the less known properties of R^2 is that it is a central measure of a random variable that is not normally distributed. Ronald Fisher, however, developed a transformation, which he called the z-transform, which allowed comparing models by creating a normally distributed random variable with mean and variance defined by the R^2 and the number of data points that were used in the regression. Most of the standard (parametric) statistical tests are designed for normal variables, which allows one to compare models based on the z-transform R^2.

Information Criteria

In the 1950s, information theory was on the rise. The concept of entropy as the measure of chaos was adopted from physics and finally put to good use by Claude Shannon, John Von Neumann, and Norbert Wiener. The approach to measurement of goodness of fit from the information-theory side is beautiful: to view regression from the standpoint of its contribution to the reduction in the chaos of our perception of the world.

From this approach, the Akaike and the Bayesian Information Criteria (AIC and BIC, respectively) emerged. A model is an approximation, and therefore, it is guaranteed to lose some information (see Fig. 2.24) about the world it describes. If we have two models, A and B, then the AIC and the BIC measure how much more information is lost by applying model A than by applying model B. Consequently, we choose the model that loses less information. The difference is in the way they account for the number of points in the independent variable that are used in building the regression.

2.5.6 Model Competition

By using the goodness-of-fit criteria and/or their combination (e.g., [GILG2010]), we can identify the model that would be the best fit for the data.

For example, consider the data shown in Fig. 2.24.

We have a quadratic relationship between the two data sets (Y_q and X). We will compare two regression lines: a straight line and quadratic parabola.

First, let us look at the straight line (Fig. 2.25).

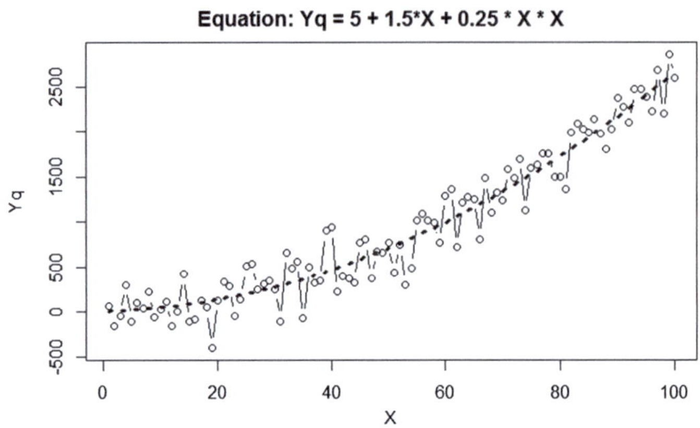

Fig. 2.24 An illustration

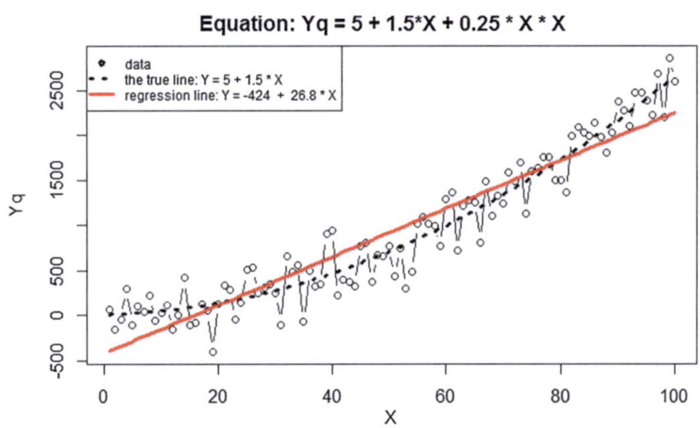

Fig. 2.25 Linear model

This line does show the direction of the correlation and fits the data fairly well; however, the $R^2 = 0.87$—not a very good fit (could be better); 13% of the variance has remained unexplained. The AIC for this model is 1420.

Next, we apply the quadratic model (Fig. 2.26).

The blue line seems to follow the data much closer. The $R^2 = 0.93$—a better fit is confirmed; only 7% of the variance has remained unexplained. The AIC for this model is 1367.

In this particular case, the R^2 and the *AIC* concur that the quadratic model does indeed provide a better fit to the data.

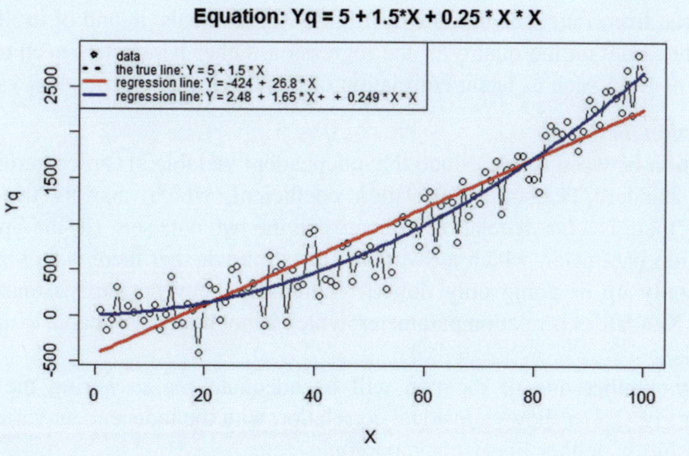

Fig. 2.26 Linear (*red*) and quadratic (*blue*) models

2.5.7 *Analysis of Residuals*

Visual

The goal of Least-Squares regression is to draw a line through the data, such that the residuals (observed−predicted) are random, with the distribution centered at zero.

In Fig. 2.26, we see that the quadratic line is a better fit than the straight line. Its residuals look as shown in Fig. 2.27a. Residuals from the straight-line regression are shown in Fig. 2.27b.

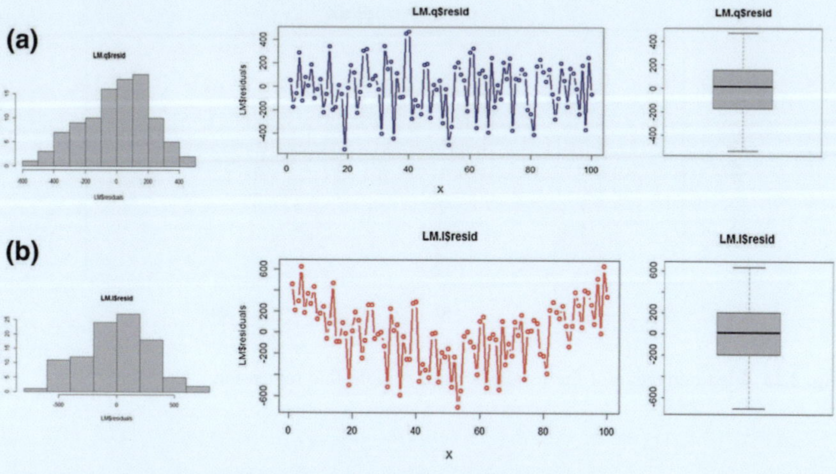

Fig. 2.27 Residuals from quadratic (**a**) and straight-line (**b**) regression

We see from Fig. 2.27 that the distribution of residuals, in and of itself, is not a sufficient signal on the quality of the regression. Other parameters need to be considered as well, such as latent correlation of residuals with independent variable(s).

Correlation Test

Correlation between residuals and the independent variable(s) can be performed via (i) the standard Pearson's correlation coefficient, which answers the question whether there is a *linear* relationship between the two data sets, (ii) the Spearman's correlation parameter, which answers the question whether there is any *monotonic* (going only up or going only down) relationship between the two data sets, or (iii) the Kendall's correlation parameter, which is not really applicable to analysis of residuals.

Sadly, neither one of the two will be adequate for answering the question whether Fig. 2.27b shows a residual correlation with the independent variable: This relationship is neither linear nor monotonic.

Yet we need to know whether the model is bad.

One way to do it is to check the correlation of the first derivative ($\frac{\Delta \text{Residuals}}{\Delta X}$) with the independent variable. If the line is quadratic, then the first derivative will be a straight line, and we will accurately capture the correlation if it is present.

This does look like a flat line; however, the formal correlation test tells us that we are still not out of the woods, the reason being the variance in the data: The signal-to-noise ratio is too low.

The other method involves splitting the independent variable into halves and running the Spearman's correlation test on both sides. If one is going up, while the other is going down, it means that residuals have a unimodal correlation.

In the scenario we are considering (Fig. 2.28), this method turned out to be beneficial. Correlation test for the first half returned.

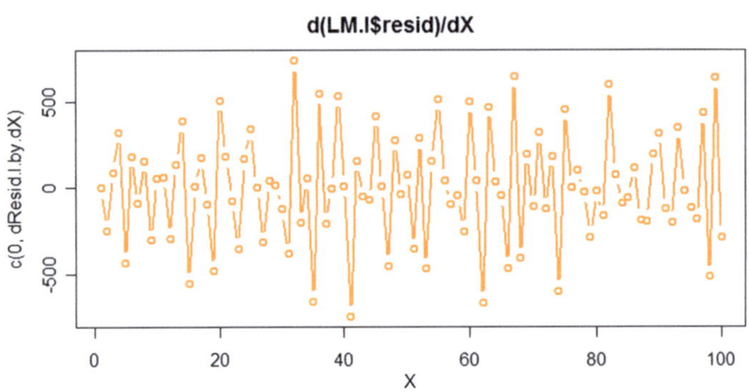

Fig. 2.28 First derivative of the residuals from straight-line regression

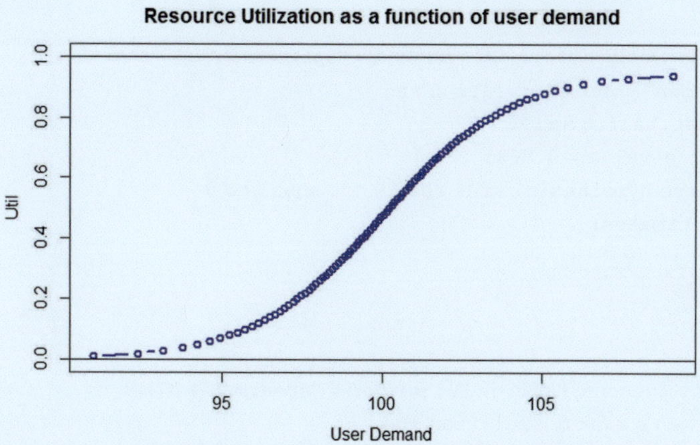

Fig. 2.29 Resource utilization is typically characterized by the sigmoid curve

```
> cor.test (first.half, X1, method = "spearman")
   Spearman's rank correlation rho
data: first.half and X1
S = 33668, p-value = 3.174e-06
alternative hypothesis: true rho is not equal to 0
sample estimates:
   rho
-0.6167107
```

For the second half:

```
> cor.test (second.half, X2, method = "spearman")
   Spearman's rank correlation rho
data: second.half and X2
S = 7212, p-value = 1.779e-07
alternative hypothesis: true rho is not equal to 0
sample estimates:
   rho
0.6736652
```

The product of the two correlation coefficients is $\rho_1 * \rho_2 = -0.6167 *0.6737 = -0.4161 < 0$.

We have correctly captured the fact that the residuals hit a low point near the middle of the range of the independent variable. By contrast, for the quadratic line, not only are the correlations not very likely (the *p*-values are high), but also the correlation parameters for both halves of the range of the independent variable are on the same side of zero:

```
> cor.test (first.half.q, X1, method = "spearman")
    Spearman's rank correlation rho
data:  first.half.q and X1
S = 20302, p-value = 0.8623
alternative hypothesis: true rho is not equal to 0
sample estimates:
     rho
0.02511405
```

```
> cor.test (second.half.q, X2, method = "spearman")
    Spearman's rank correlation rho
data:  second.half.q and X2
S = 21068, p-value = 0.7442
alternative hypothesis: true rho is not equal to 0
sample estimates:
     rho
0.04669683
```

```
> cor(first.half.q, X1) * cor (second.half.q, X2)
[1] 0.001776423
```

Based on this analysis, despite the preference we should always give to the more simple models, given the choice between straight-line and quadratic relationships, we choose the quadratic model for the data in Fig. 2.27.

2.5.8 Advanced Regression Methods

Detailed coverage of advanced topics on regression is outside the scope of this book. Interested readers are referred to specialized texts covering in detail nonlinear regression; logistic regression; the general linear model (GLM), and other methods of regression. Majority of statistical and data mining software products have libraries that implement such methods. However, we will say a few words here about each of them.

Nonlinear Regression
Nonlinear regression is used when we know the form of the equation, and it is not one of the "Big 6." In such cases, iterative methods are used to compute the coefficients.

Generalized Linear Model (GLM)

The GLM is used for a wide variety of scenarios where the distribution of the residuals cannot be normal (Gaussian). In such cases, the GLM still tries to draw a line through the mean of the distribution; however, the mean will not necessarily correspond to the central point in the distribution.

Logistic Regression

A very special, and widely used, case of GLM is the logistic regression. It is so prominent in predictive analytics that most statistical software products have it implemented as a separate method. Logistic regression is used when we want to determine the probability [or any other variable bound between two hard limits (typically 0 and 1, but that can be scaled to any other pair of limits) as a function of some independent variable(s)]. In the IT world in general, and in LTE networks in particular, utilization (e.g., CPU, memory, bandwidth) falls into that category: it can be viewed as the probability that at any moment in time the system will be in the busy state, which makes it important to discuss logistic regression in this book.

We will also use it to illustrate the need to reverse all transformations and substitutions that were done when linearizing the regression equations.

Utilization of a live (operational) resource can never be greater than 1 and less than 0, but it can approach these limits asymptotically. This imposes a very distinct shape on the curve expressing utilization as the function of, e.g., user demand (Fig. 2.26).

The sigmoid (aka logistic) curve is the 7th of the "Big 6" linearizable curves. Indeed, if we consider that utilization is probability of being busy, we can compute the odds of being busy:

$$\mathrm{Odds_{busy}} = \frac{\mathrm{Util}}{1 - \mathrm{Util}} \tag{2.5.3}$$

That, however, will result in an exponential-looking curve (Fig. 2.30).

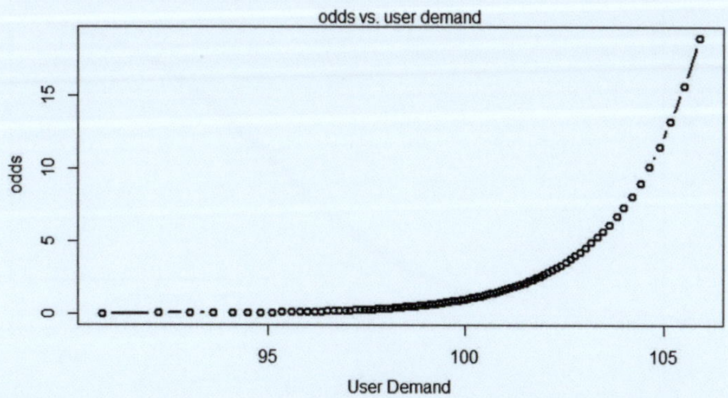

Fig. 2.30 Odds of seeing the resource in a busy state as a function of user demand

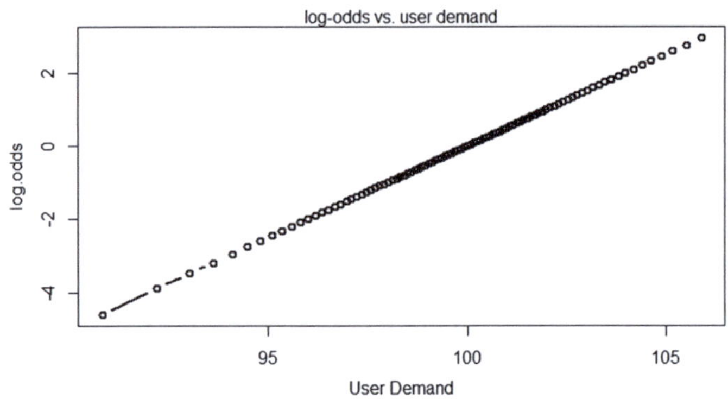

Fig. 2.31 Log-Odds of seeing the resource in a busy state as a function of user demand

One final operation—taking the natural logarithm of the odds—will result in a straight line (Fig. 2.31):

$$\text{log-Odds}_{\text{busy}} = \ln\left(\text{Odds}_{\text{busy}}\right) = \ln\frac{\text{Util}}{1 - \text{Util}} \tag{2.5.4}$$

One final step is to reverse the calculations to go from linear model prediction for $\text{log-Odds}_{\text{busy}}$ to Util in Fig. 2.29.

$$\text{prob}_{\text{busy}} = \frac{e^{\text{log}-\text{Odds}_{\text{busy}}}}{1 + e^{\text{log}-\text{Odds}_{\text{busy}}}} \tag{2.5.5}$$

Equation (2.5.5.) was derived from Eq. (2.5.4) by simple algebraic transformations (Fig. 2.32).

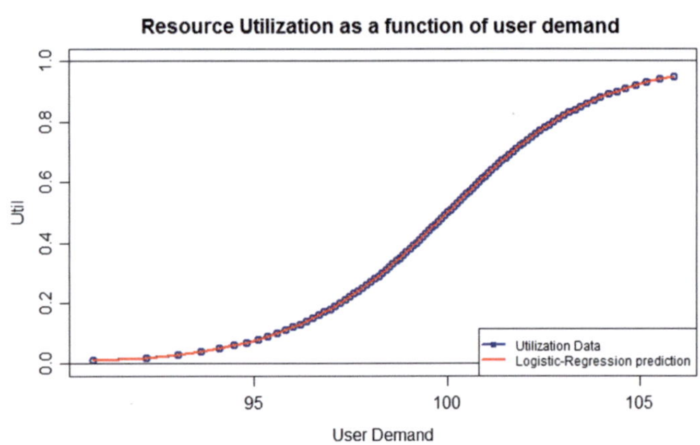

Fig. 2.32 The final regression fit

In performance and fault analysis, sometimes the prediction line ends up above the data. This happens as an artifact of applying Eq. (2.5.5): In effect, we are scaling the range of utilization to the [0, 1] range. If it happens, it is a strong indicator of hidden saturation (congestion) on a related resource.

Quantile Regression

The quantile regression [KNKR2001] approach is relatively new. It was first formulated in 1987; however, the computing power was not up to par, making the software implementing QR sluggish at best. Fifteen years later, the quantreg package was implemented for *R*, and another few years after that, quantile regression was implemented in SAS. Today, it is widely used everywhere, especially in social sciences, where families in the socioeconomic standings react differently to different stimuli. It is useful in building predictive models for LTE networks, e.g., where technically different user elements share a bandwidth. This will lead to a variety of usage patterns, some of which can even be used to identify the user's equipment.

Quantile regression is highly efficient when we do not know, and cannot guess, the type of distribution of the dependent variable.

It can also be used to dynamically detect outliers in correlated data: Indeed, if we can draw regression lines through the 25th and 75th percentiles (first and third quartiles) of the dependent variable, compute the IQR (see section on Outliers in this chapter) at every "vertical" slice and compute the trajectories of high and low whiskers (high and low outlier boundaries).

Other examples of using quantile regression include capacity planning, busy-hour-traffic analysis, and Predictive SPC [FERR2014].

2.5.9 Do We Have to Compete?

Sometimes, it is hard to choose the "right" model, even when formally the parameters are pointing at one. When formal competition yields a tie, it becomes even more complicated. When that happens, we need to either choose the model based on some heuristics or use the so-called *ensemble approach*.

With ensemble, or Delphi (in honor of the Delphi Oracle in Ancient Greece) approach, we combine predictions of two or more models into one. For some models, it can be easily done by simple formulae; for others, more advance techniques, such as bootstrapping and jackknifing, have to be used.

Ordinary Least-Squares (OLS) Regression

As we discussed, OLS regression's residuals are expected to be normally distributed and centered at zero. This provides an easy way to compute the distribution of the ensemble. Indeed, since all models in the ensemble are independent, and so are their predictions, we can compute the mean at all values of independent variable (s) as the sum of means, and the variances of the ensemble prediction as the sum of variances, of the models that were used in the ensemble:

$$\forall \boldsymbol{X}: \quad \mu_X = \sum_{m=1}^{M} (\mu_X)_m \tag{2.5.6a}$$

And

$$\forall \boldsymbol{X}: \quad \mathrm{Var}_X = \sum_{m=1}^{M} (\mathrm{Var}_X)_m \tag{2.5.6b}$$

In Eqs. (2.5.6a) and (2.5.6b),

X	is the independent variable (or a combination of covariates)
M	is the number of models in the ensemble
$\mathrm{Var}_X = \sigma^2$	is the variance of residuals predicted at $\boldsymbol{X}$

Because (see Section 2.5.7) we do not want to use a model that leaves residuals cor-related with X; Eq. (2.5.6b) can be simplified to:

$$\forall \boldsymbol{X}: \quad \mathrm{Var}_X = \sum_{m=1}^{M} \mathrm{Var}_m \tag{2.5.6c}$$

Note: Model predictions will appear to be correlated. This is expected: We are modeling the same process using different models. However, the models are independent; therefore, Eqs. (2.5.3a–c) will work.

Non-OLS Regression
For non-OLS regression, when the distribution of residuals is generally non-normal (see Sects. 2.1–2.3), ensemble methods will involve more complicated statistical techniques, falling under the general classification as Monte Carlo analysis. They are outside the scope of this book. A simple, down-to-earth, technique that allows the analyst to avoid using Monte Carlo and yet get decent results involves resampling. The idea is to fulfill the requirements of the Central Limit Theorem: take multiple independent samples at random from the values predicted by each model. If enough such samples are taken, their means will converge to a normal distribution, and then, Eqs. (2.5.6a)–(2.5.6c) will be usable for computing the ensemble prediction.

2.6 Clustering

In LTE analytics, it is often important to break data into natural groups.

One of the most immediate use cases involves bundling curves into clusters. Another use case is dealing with geographic clustering (e.g., identifying natural groupings of UEs to determine RAN metro boundaries).

We shall start with the first use case, as it builds on the material we introduced in this chapter in Sects. 2.4 and 2.5.

2.6.1 Bundling the Curves

In Fig. 2.33, we show a group of data simulated to reflect 500 days of bandwidth utilization data for five network links.

We see interesting patterns in data: In the early days (May through August, 2010), the five links were ramping up together, reached what appears like saturation at ~90%. This was followed by a drop in utilization of L4 and L2 (perhaps they were taken out of traffic for a few months). The other three links (L0, L1, and L3) saw a sharp drop in utilization around the same time. Such drops are usually associated with an increase in capacity.

L1 started ramping up gradually, while the other two (L1 and L3) stayed at the same level of utilization to which they came after the drop.

Finally, in February 2011, links L2 and L4 were put back into operation.

This is all a very basic analysis that can be performed visually, but when dealing with hundreds, even thousands, of links, we want an automatic system to perform it for us. We need a mathematically soun way to automatically group data into bundles.

K-means

The K-means method of clustering is the simplest yet surprisingly powerful. *K* "Centers" are put at random on the field (*y*-axis), and all data points are assigned to each of the *K* clusters so as to minimize their distance to the center. The new

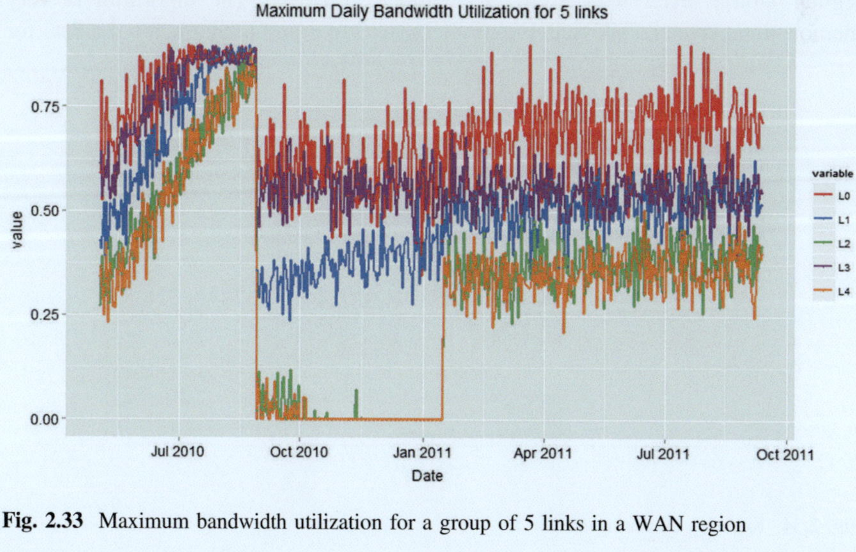

Fig. 2.33 Maximum bandwidth utilization for a group of 5 links in a WAN region

centroid (mean) is calculated for each cluster, and the data points are reassigned. The iterations continue until no more improvements to the distances to the centroids can be made. Due to the way distances are calculated (typically sum of squares of Euclidean distances), this method is equivalent to moving K variable-width Gaussian bell curves (surfaces if the clustering is multi-dimensional) until the narrowest possible bell curves are achieved.

Applied to the data shown in Fig. 2.33, K-means clustering produces decent results, which, however, fail to detect the patterns in data trajectories.

Advantages of K-means Clustering
It is simple, powerful, and versatile.

Downsides of K-means
K-means method groups data based on their value, without consideration for trending and other patterns in data. Consequently, cluster overlaps are to be expected when this method is applied to time-series data. We see these overlaps in Fig. 2.34.

The clustering shown in Fig. 2.34 better reflects what we think we see from the data, but both groupings are wrong, and both groupings are right.

Model-Based Clustering
Model-based clustering (commonly abbreviated as MBC) is an extension of K-means for multivariate clustering, where the bell surfaces (Gaussian clouds) are extended along the primary axes of the data. For technical details, please see [FRAF2002] and [FRMS2012].

It is a highly sophisticated algorithm, fitting different shapes of Gaussian clouds into data and using the BIC (see Sect. 2.5) to decide when to stop adding clusters (detect the point of diminishing returns in information gain).

The particular data set that we have been looking at is too big for MBC on a regular laptop, even with 8 GB of memory, however. The algorithm is very memory-intensive. To alleviate the problem, we are going to aggregate the data for

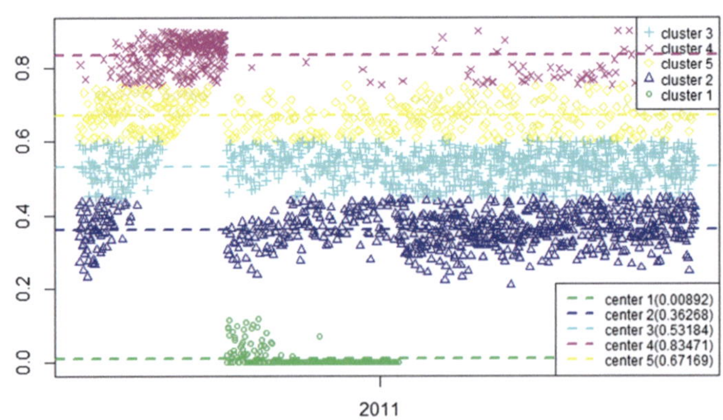

Fig. 2.34 K-means clustering at $K = 5$

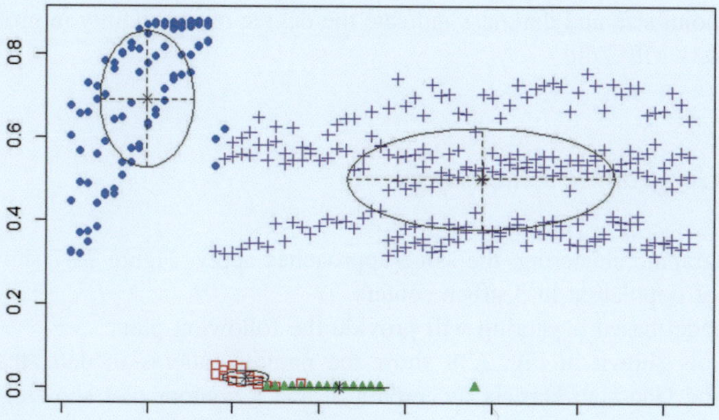

Fig. 2.35 Clustering by model-based method

each set into weekly averages. The reason we choose averages is because MBC relies on a Gaussian Mix Model; so using averages actually helps in this case by fulfilling the conditions of the Central Limit Theorem. Doing so also reduces the data by a factor of 7. It still takes a couple of minutes, but the result is very informative (Fig. 2.35).

The ellipses that MBC method has added show the Gaussian kernels that characterize the clusters. The colors correspond to the clusters. We see how different sets from Fig. 2.33 group naturally with other sets at different times, which allows us to see the dynamics of the data. This allows us to build an automatic performance monitoring system that will answer important performance-related questions.

The classification uncertainty plot (Fig. 2.36) allows the analyst to see regions where the model-based classifier had higher and lower confidence of the cluster assignment.

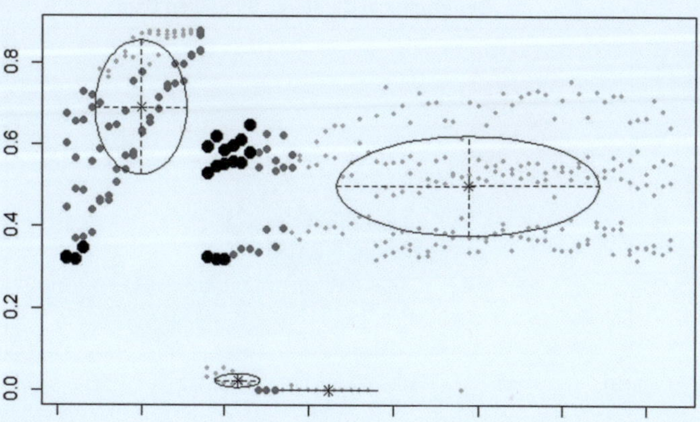

Fig. 2.36 Classification uncertainty chart

Here, point size and darkness indicate the degree of uncertainty in grouping the data the way MBC did.

2.6.2 Geographic Clustering

With geographic clustering, the same approaches apply. Figure 2.37 shows a distribution of population in 3 urban centers.

The model-based clustering will provide the following plots.

The plots shown in Fig. 2.38 show the natural patterns in data, the overlap between the Gaussian kernels (uncertainty), and a contour plot showing the data density distribution. The cluster assignment (count and shape) is optimized based on the Bayesian Information Criterion (BIC). The process is illustrated in Fig. 2.39.

Here, the 3-letter combinations correspond to the shapes of the cluster kernels, and the term "components" corresponds to the clusters into which the data is grouped. For more details, please refer to the literature provided below.

2.6.3 Geographic Clustering of Signal

The method we have illustrated here assigns clusters based on the density of the points on the map (see Fig. 2.39). When we are dealing with signal distribution within each cluster, a different method has to be used, the DBSCAN. Its biggest advantage over the K-means and the MBC methods is that it allows identifying groupings of data with nonlinear separation boundaries. Its discussion is outside the scope of this chapter. R, python, SAS, SPSS, and other statistical tools all have

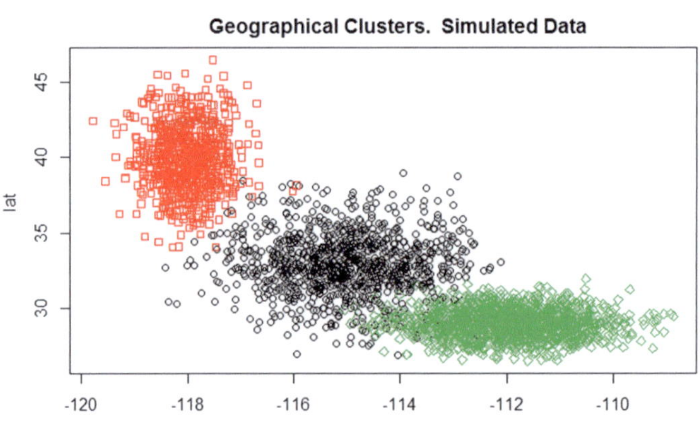

Fig. 2.37 The three urban centers

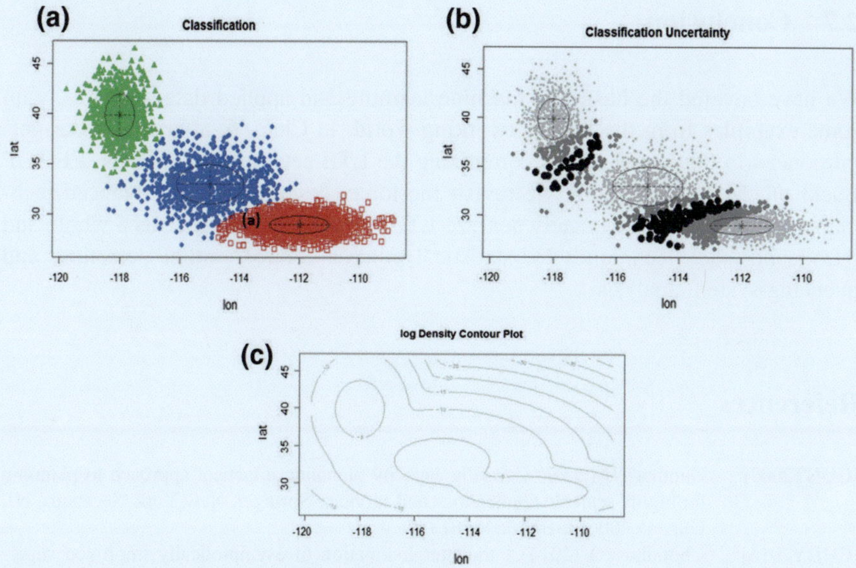

Fig. 2.38 The three plots provided by the model-based clustering

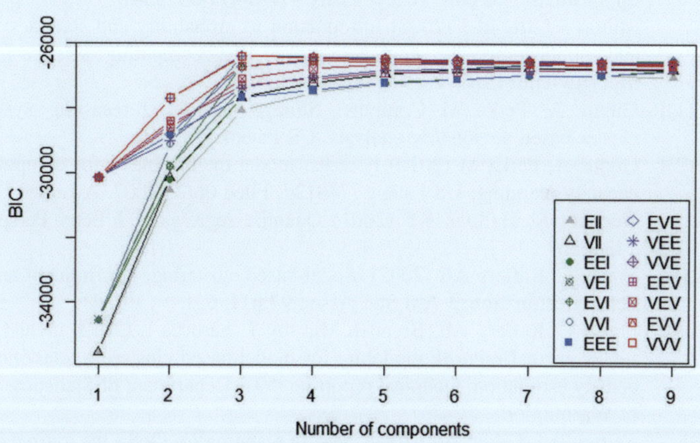

Fig. 2.39 An illustration of the cluster assignment process

implementations of K-means, MBC, DBSCAN, and other separation methods shipped either as part of the standard distributions or as additional libraries.

We have only touched upon the most basic use cases and tools for clustering. For more information, techniques, and tools, please see the literature referenced in this book.

2.7 Conclusion

We have covered the basics of machine learning and applied data analytics, with some examples from the LTE networking world. In Chap. 6, after the reader gets introduced to the key elements comprising the LTE network—EUTRAN, EPC, IP Backhaul, Metro, and Core, we revisit the topic, describing why advanced techniques are needed to adequately analyze LTE network performance as a whole and introducing such techniques for statistical process control, outlier detection, and queueing-system analysis.

References

[GUNT2007] Gunther NJ (2007) Guerrilla capacity planning: a tactical approach to planning for highly scalable applications and services. Springer, New York, Secaucus, NJ, USA © 2006. ISBN 3540261389

[CHDY2014] Choudhury J (2014) Parameter Estimation of asymptotically improved super-serial scalability law. In: Performance and capacity international conference by computer measurement group (CMG'14) (PDF)

[MAKR1998] Makridakis S, Wheelright SC, Hyndman RJ (1998) Forecasting: methods and applications, 3rd edn. Wiley. ISBN 978-0-471-53233-0

[MILT2002] Milton S, Arnold J (2002) Introduction to probability and statistics: principles and applications for engineering and the computing sciences, 4th edn. McGraw-Hill. ISBN-13 978-0072468366

[GILG2010(1)] Gilgur A, Perka M Computer Storage capacity forecasting system using cluster-based seasonality analysis. US Patent 7783510

[GILG2010(2)] Gilgur A, Perka M (2010) Forecast model quality index for computer storage capacity planning. US Patent 7788127. Filed 06/25/2007. Awarded 08/24/2010

[KNKR2001] Koenker R, Hallock KF (2001) Quantile regression. J Econ Perspect 15(4): 143–156

[FRAF2002] Fraley C, Raftery AE (2002) Model-based clustering, discriminant analysis and density estimation. J Am Stat Assoc 97:611–631

[FRMS2012] Fraley C, Raftery AE, Brendan Murphy T, Scrucca L (2012) mclust Version 4 for R: normal mixture modeling for model-based clustering, classification, and density estimation, technical report no. 597, Department of Statistics, University of Washington

[FERR2014] Ferrandiz J, Gilgur A (2014) Capacity planning for QoS - the journal of capacity management. A publication of the computer measurement group. Issue 135, Winter, 2014

Chapter 3
Evolved Universal Terrestrial Radio Access Network (EUTRAN)

Mobile wireless network has evolved from the first generation of analog communications to the fourth-generation evolved universal terrestrial radio access network (E-UTRAN). E-UTRAN access technology is also referred to long-term evolution (LTE) with its spectral efficiency increased 150 times as compared to the first-generation analog access technology.

In this chapter, we will discuss RAN architecture in Sect. 3.1, followed by LTE air interface specifications in Sect. 3.2, evolved Node B base station in Sect. 3.3, then present RAN network management model in Sect. 3.4, and finish with RAN network major KPI and analyses in Sect. 3.5. Chapter summary is given in Sect. 3.6.

3.1 RAN Architecture

LTE radio network architecture supports multiple frequency bands, various radio nodes, and versatile of services. Radio network efficiency has dramatically improved over a much wide bandwidth.

Figure 3.1 illustrates LTE–RAN architecture. LTE is the first access interface standard that has defined heterogeneous network features for a radio network involved in macro, micro, femto, and other technologies, such as Wi-fi, CDMA2000, and HSPA. The macro- and micro-base stations are connected with EPC through control plane S1-MME interface and user plane S1-U interface.

Radio network with radio and baseband in one location or centralized RAN with distributed radio nodes and centralized baseband processing. It supports both traditional base stations with dedicated backhaul and Home eNB (femto) with public backhaul through home Internet connection.

D. Kakadia et al., *Network Performance and Fault Analytics for LTE Wireless Service Providers*, DOI 10.1007/978-81-322-3721-1_3

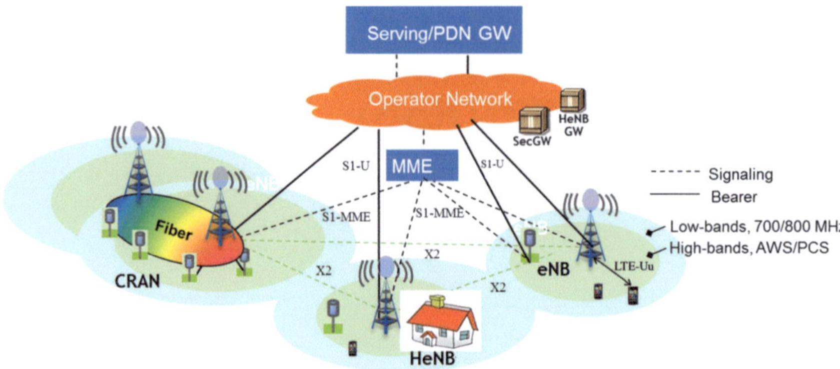

Fig. 3.1 LTE RAN architecture

3.1.1 Evolved Node B

Evolved Node B (eNB) performs a traditional base station's functions of transmitting and receiving radio link signals. It is the radio link termination point for both forward and reverse link physical (PHY), medium access control (MAC), radio link control (RLC), and packet data convergence protocol (PDCP) layers. eNB performs H-ARQ, fast power and rate control, connected mode mobility, and radio resource management of various QoS. Header compression (such as Robust Header Compression—ROHC) and ciphering are also implemented at eNB for enhanced security.

Figure 3.2 shows the functionality splitting among eNB, mobility management entity (MME), serving gateway (SGW), and packet data network gateway (PDN-GW or PGW). The splitting of user and control plane enables flexible network layout. This could allow more centralized MME. SGW can be more closed to,

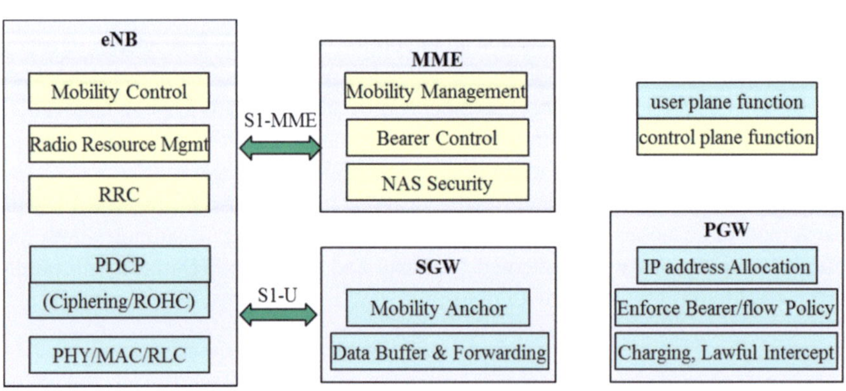

Fig. 3.2 Function splitting among eNB, MME, SGW, and PGW

where traffic is generated with reduced latency and transport. Distributing intelligence among the base stations speeds up the connection setup and reduces the time required for a handover.

Various categories of eNBs are utilized to efficiently densify network and address exponential traffic growth. Traditional macro-eNB with transmission power above 10 W has been expanded to micro-cell with power around 5 W, pico-cell power around 1 W, and indoor radio nodes of 50 mW. Heterogeneous network features are defined in LTE-advanced cells to coordinate various cells, such as further enhanced inter-cell interference coordination (feICIC).

3.1.2 Centralized RAN

Cell densification demands tight coordination among radiated radio nodes. Centralized RAN consists of centralized high-capacity baseband processor and distributed radio nodes. It separates remote radio head (RRH) from the digital baseband unit (BBU) by dark fiber through CPRI standard interface.

CRAN enables operational flexibility and efficiency by providing cost-saving on real-estate, utility, and eNB maintenance. It enables capital saving on resource sharing and pooling among various nodes. It can coordinate radiated power, time, and frequency resources among radio nodes to achieve better mobility and radio capacity. It can also utilize LTE-advanced features such as coordinated multiple point (CoMP) transmission and reception features, and carrier aggregations among frequencies supported by various nodes.

Figure 3.3 illustrates cell site traffic variation inside and outside of a stadium during a game. The variation shows potential traffic pooling gain and resource saving.

CRAN architecture requires higher bandwidth and low latency compared to traditional Ethernet backhaul. This limits the distance within 9 miles between the

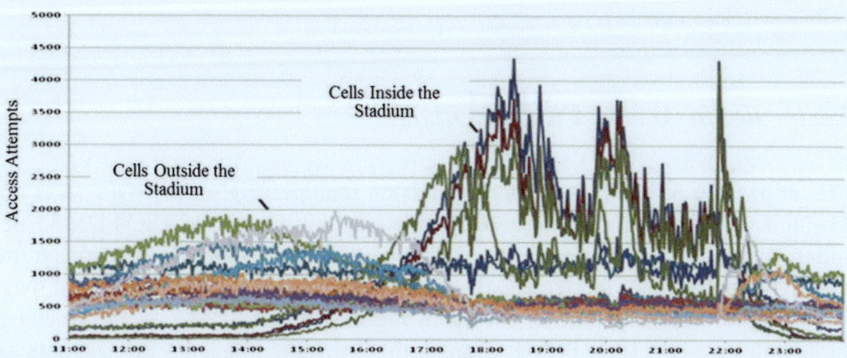

Fig. 3.3 Traffic pooling and resource saving using CRAN

remote nodes and the central BBU. As backhaul cost difference diminishes with large radio bandwidth and availability of dedicated dark fiber, CRAN has become one of the viable options for eNB evolution.

3.1.3 Home eNB

A home eNodeB or HeNB is the 3GPPs term for a LTE femto cell, privately owned low-power unit. HeNB performs the same function as that of an eNB, but it is connected to operator network through a public network. It is typically used to expand the coverage or capacity of an indoor environment, where the wireline cable or fiber is readily available. The HeNB is connected with a security gateway (SeGW) installed in an operator's network. The SeGW establishes IPsec tunnels with HeNBs using IKEv2 signaling for IPsec tunnel management. IPsec tunnels are responsible for delivering all voice, messaging, and packet data services between the HeNB and the core network. HeNG gateway (HeNB-GW) could also be presented between HeNB and MME for S1-MME control plane interface and between HeNB and SGW for S1-U user interface.

Home eNB is an efficiency way to densify network together with other in-building solutions, macro-, micro-, and pico-cells. It is an essential part of capacity solutions to address traffic growth.

3.2 Air Interface Specifications

LTE air interface is designed to achieve high spectral efficiency, high peak rate, short round trip time, and flexibility in frequency and bandwidth. The access network has achieved the spectral efficiency close to Shannon's limit under wide range of mobile radio conditions. Physical and logical channel structures support dynamical channel allocation and radio resource management for a variety of services. LTE is established as the preeminent access technology for wide area mobile wireless network.

3.2.1 Radio Access Fundamentals

LTE utilizes an orthogonal frequency division multiplexing access (OFDMA) on downlink and single-carrier frequency division multiple access (SC-FDMA) on uplink. This is a multicarrier access technique that divides the available spectrum into many narrow and orthogonal sub-carriers. The information is transmitted over sub-carriers on discrete time as shown in Fig. 3.4. The OFDMA is implemented through inverse FFT (Fast Fourier Transform) to transfer frequency domain data to time domain signal. The only difference in SC-FDMA on uplink is one extra

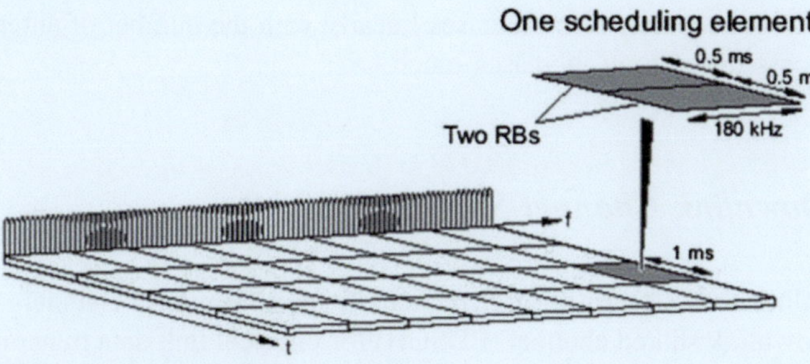

Fig. 3.4 LTE radio access structure

discrete Fourier transform to mitigate the peak-to-average power ratio problem on user device.

LTE has much finer granularity on both time and frequency domains compared to previous generations of air interface standards. Schedulable transmission time interval (TTI) is only 1 ms, versus original schedulable frame length 20 ms in CDMA2000-1x or 10 ms in WCDMA network. Each resource block (RB) is 180 kHz, much higher resolution than that of 1.2288 MHz CDMA2000-1x or 5 MHz UMTS bandwidth. Current 3GPP Rel.8 to Rel. 12 has defined channel quality indicator (CQI) reported by user device per TTI. It can be reported either over every sub-band (6 RBs) or wide band (entire bandwidth). Resource can be scheduled over each RB per TTI. This significantly improves the network efficiency through time- and frequency-selective scheduler.

The OFDMA access technology has many advantages. OFDMA-based solution is significantly more efficient than that of CDMA-based solution against adverse multipath fading radio environment, in particular, over large bandwidth spectrum. OFDM access scheme can also easily accomplish multiple inputs and multiple outputs (MIMO) antenna structure. MIMO is required to achieve high peak and average user throughputs for broadband communications. Receiver complexity for MIMO-CDMA grows exponentially with bandwidth and linearly with MIMO-OFDM. Figure 3.5 illustrates OFDMA implementation over multiple-element transmitter and receiver.

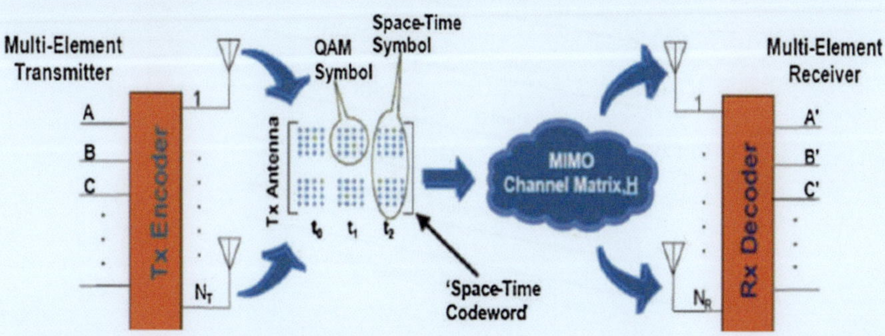

Fig. 3.5 Multiple inputs and multiple output antenna elements in LTE

The OFDMA capacity also increases linearly with the number of antenna, while CDMA access technology increases much less.

3.2.2 Downlink Channel Structure

LTE downlink (base station to user device direction) physical channels consist of physical downlink shared channel (PDSCH) for physical link data transmission and various control channels as shown in Fig. 3.6.

Synchronization channels (SCH) carry primary synchronization signal (PSS) and secondary synchronization signal (SSS) for user devices to discover the LTE cell and initial synchronization. Physical broadcast channel (PBCH) broadcasts basic system information within a cell. Physical downlink control channel (PDCCH) carries downlink allocation information and uplink allocation grants. Physical control format indicator channel (PCFICH) signals the length of PDCCH. Physical hybrid ARQ indicator channel (PHICH) carries the acknowledgment of uplink transmission.

3.2.3 Uplink Channel Structure

The uplink communications (from user device to base station) are illustrated in Fig. 3.7. Physical uplink shared channel (PUSCH) carries the physical layer's uplink data transmission together with control information. It utilizes modulation formats QPSK, 16 QAM, and 64 QAM. Physical random access channel (PRACH) is used by user device for initial access or when device losses its uplink connection.

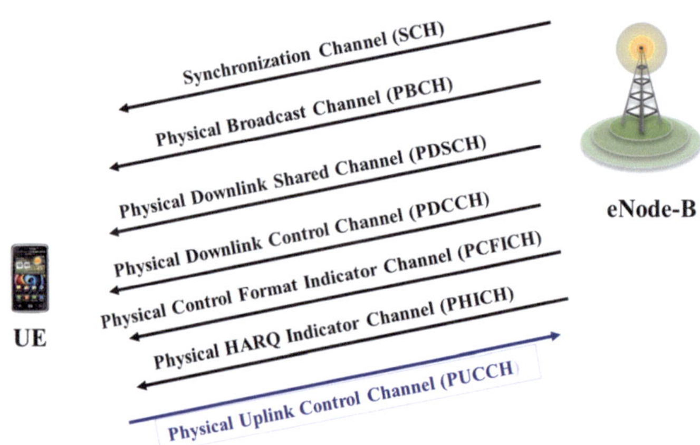

Fig. 3.6 Physical channels to support downlink communications

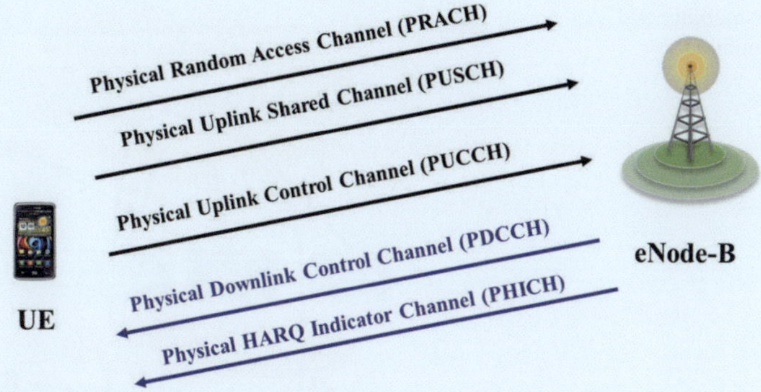

Fig. 3.7 Physical channels to support uplink

Physical uplink control channel (PUCCH) transmits DL acknowledges and CQI-related reports. Uplink coding and allocation grants are determined by eNB and signaled to the user device in the PDCCH.

When a user powers up a mobile device, it will listen to SCH and PBCH to get system information. Then, it will accesses LTE network through PRACH channel. After receiving acknowledgment from PDCCH, it will transfer data over PDSCH or PUSCH data channels accordingly.

Downlink transmission depends on uplink PUCCH for hybrid ARQ acknowledges and CQI feedback to estimate channel distortion. Uplink transmission relies on resource allocation grant sent over downlink PDCCH and hybrid ARQ acknowledges over PHICH. Those channels interact with each other. A successful LTE connection requires proper connections of the data and control channels on both downlink and uplink. As the number of user increases, efficient resource allocation and real-time control among time, frequency, and power domains become even important to ensure a properly operational access network.

3.2.4 LTE Mobility

Mobility management is essential for a wide area wireless network. Separation of user plane and control plane requires involvement and quick response among MME and eNB to ensure robust and fast handovers.

Two layers of mobility are defined in LTE, IP mobility, and radio link mobility, see Fig. 3.8. IP mobility supports fast and efficient IP handover by maintaining the same IP address. This ensures session's continuality and seamless user experience. LTE radio link handover is initiated by source eNB with no macro-diversity. Both S1 interface- and X2 interface-based handovers are defined and supported by eNBs. Direct data forwarding can be enabled among eNBs for delay-tolerant applications. This allows lossless and short interruption time. Radio resource control

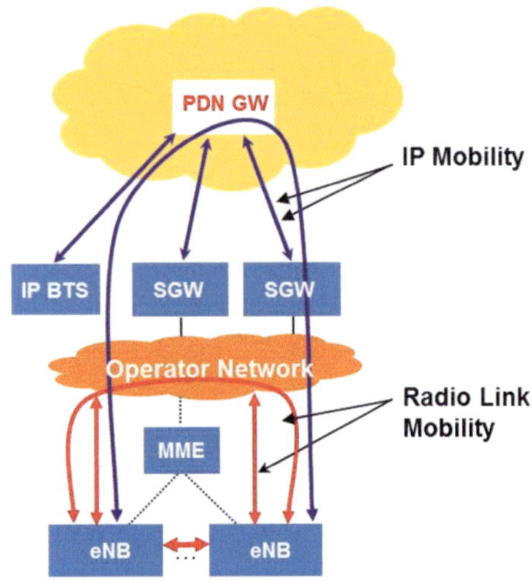

Fig. 3.8 IP mobility and radio link mobility

(RRC) connection reestablishment during handover failure is also defined. The target eNB can fetch the user device context from source eNB to recovery from radio link failure and reduce the potential connection loss during handover.

LTE dynamic channel structure and flexible resource management can efficiently support various services and network configuration. Operator can deploy a heterogeneous network to address exponential traffic growth. However, sophistic network management and self-organizing network will be required to effectively operate this network.

3.3 Evolved Node B—Base Station

LTE base station has built in more software definable radio (SDR) features than traditional base station. High capacity requires more advanced parallel processing. High speed demands large data buffer inside the base station modem. Multiple transmitting and receiving elements need multiple radio chains. Intelligent scheduler ensures an efficient resource allocation.

A generic base station consists of transceiver, baseband processing, control processor, memory, and various input/output components, as illustrated in Fig. 3.9. LTE transceiver cards have power amplifier (PA) and low-noise amplifier (LNA), digital intermediate frequency (IF), and various RF filters. They are connected to antennas on the one side and to baseband channel cards on the other side.

Channel cards or modems perform physical channel and MAC channel processing, multiplex and de-multiplex as well as interference cancellation and scheduler functionalities.

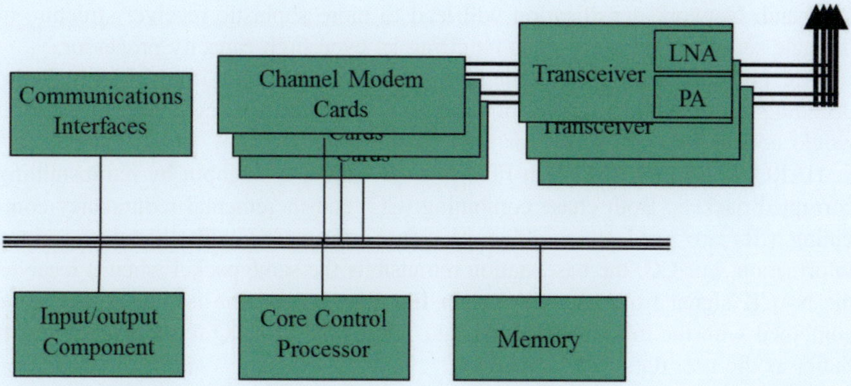

Fig. 3.9 Base station architecture

The capability and robustness of a base station directly impact performance and reliability of a wireless network. For example, the main processor capacity could limit the number of simultaneous users a base station supports. Transceiver design will determine the number of multiple antennas and transmission power in a cell. The memory size could impact the latency and throughput. Therefore, it is critical to carefully monitor main processing and memory utilizations in an eNB, optimally control base station transmit power, and mitigate interference.

3.3.1 Baseband Processing

LTE baseband processing is the heart of a base station. It can utilize ASIC, FPGA, or programmable processors. Programmable hardware has been a dominant solution to support multiple radio access technologies and enable flexible configuration of various spectrums—TDD, FDD, and half duplex.

The baseband processing includes time and frequency synchronization, IFFT/FFT, channel estimation, MIMO detection, link adaptation, HARQ and forward error correction (FEC), and subcarrier de-mapping.

Link adaptation determines modulation and coding schemes (MCS) based on CQI, precoding matrix indicator (PMI), and rank indicator (RI) feedback from user device. LTE supports five modulation schemes on various channels: BPSK, QPSK, 16 QAM, 64 QAM, and 256 QAM. It also supports different channel codings over different channels, including Turbo coding, convolution coding, repetition code, and block code.

MIMO receiver implementation is very demanding. Maximal likelihood (ML) detector is an optimum detector but too complex to implement. A simplified ML with reduced complexity and parallel processing structure is typically used at baseband. Network densification will lead to more sophistic receiver structure to mitigate severe interference, thus requiring an even high-capacity processor.

FEC codes such as Turbo code and convolution code are used in LTE. Turbo decoding required large amount of operations with tight latency constrains; thus, it would need a powerful processor with accelerators.

HARQ is used together with FEC to improve the throughput by retransmitting corrupted packets. Both chase combining (CC) and incremental redundancy combining (IR) are used to combine previous information with the retransmitted information. For CC, the base station retransmits the same packet when it receives the NACK signal from user device. In IR, new information is transmitted to be combined with the initially received data. Therefore, H-ARQ requires a large soft buffer as the user data rate increases.

Baseband processor and memory size will limit the number of simultaneous users. This user capacity could be reduced further as the user data rate and number of MIMO transmission increases.

3.3.2 *LTE Scheduler*

LTE scheduler is the brain of a LTE network. It dynamically allocates radio resource to multiple users and multiple applications at any instantaneous time. Scheduler is a key element where vendors can differentiate themselves. It constantly interacts with other radio resource management functions, such as link adaptation, power and rate control, and HARQ.

Two major categories of scheduler are implemented in LTE: One is dynamic scheduler, arming to maximize data throughputs; and the other is QoS aware scheduler, ensuring differentiated user experience for specific service. The scheduling diagram is illustrated in Fig. 3.10.

The dynamic scheduler is based on radio channel quality, user usage, and previous user data rate. For example, DL scheduling is based on CQI, PMI, and RI reported by user device. UL scheduling takes into account sounding reference

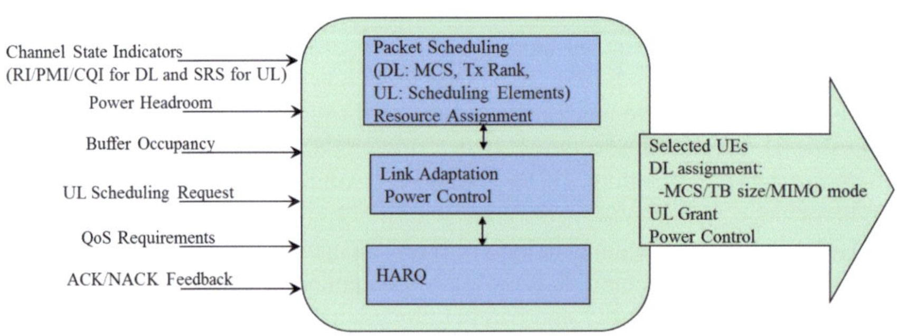

Fig. 3.10 LTE scheduler diagram

signal (SRS) and power headroom at device. Operator can configure schemes to maximize cell capacity, equal time (resource fair), equal data rate, or a balance among cell and user throughput such as Proportional Fair scheduler. The Proportional Fair scheduler can improve spectral efficiency by transmitting when instantaneous SINR is above previous average SINR level.

The QoS aware scheduler takes into account consideration priority and weighting of QCI in the case of multiple bearers with different QCIs for the same user or different users. Delay-based scheduler is one kind of QoS aware scheduler. It can increase the priority level from low to high as buffer time in the scheduler approaching the packet delay budget. This enables the bundling of several small packets to be transmitted using just one PDCCH message. Semi-persistent scheduler (SPS) is another way to resolve potential limitation of PDCCH capacity by utilizing a dynamic grant sent to semi-persistent capable device for MCS and RB assignment before transmission.

Dominant scheduler is a dynamic scheduler with Proportional Fair algorithm. Delay-based scheduler and SPS have been used for voice over LTE (VoLTE) application. With advanced performance analytics, we can further enhance the scheduler to adapt to various applications, taking advantage of end-to-end performance to maximize radio network efficiency and operation effectiveness.

LTE base station offers a wide range of opportunities to optimize itself through performance and fault analytics. The processor and buffer inside modem, as well as algorithms, could be dynamically tuned to various applications and services. Scheduler and resource management interacting with link adaptation and HARQ are intelligently adjusted to maximize network capacity, coverage, and user experience.

3.4 RAN Network Management Model

A public land mobile network (PLMN) operation highly depends on radio access network. The core management of a PLMN includes providing services to its customer, creating provision of a network infrastructure to fulfill the services, and assures the service meeting the customer requirement and bill customer for the service.

Performance and fault management are essential to assume service meeting the customer requirements. RAN network performance and fault managements follow 3GPP-defined architecture and OSS interfaces. The management model is illustrated in Fig. 3.11, with an emphasis on interfaces between the element manager (EM) and network manager (NM) of a single PLMN provider, while the element managers could come from different manufacturers to manage its own network element (NE).

The network element component, e.g., eNB, is responsible for collecting performance and fault measurements, usage data and generating alarms. The network element can contain the policy enforcement or policy decision functions.

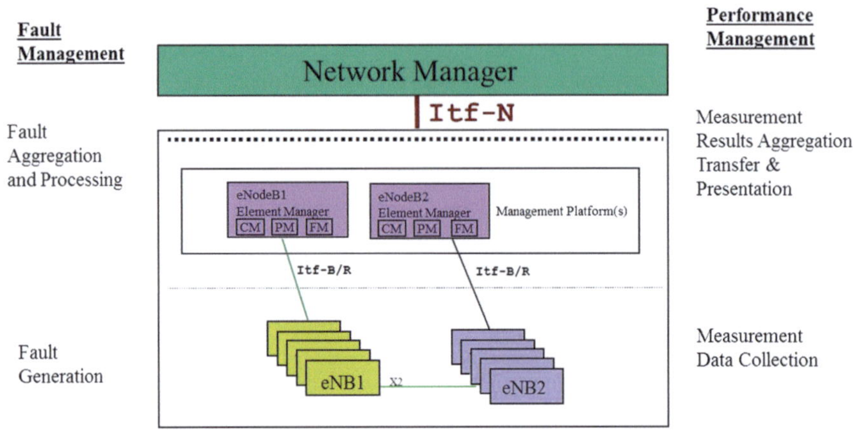

Fig. 3.11 E-UTRAN management model

On fault management side, the network elements detect fault, generate alarms, clear alarms, forward and filter alarm, store and retrieve alarms in/from the NE, recover fault, and configure alarms. The element management layer is responsible for aggregating and transferring the collected information and generated alarms/events.

On performance management side, the network elements provide measurements corresponding to various radio network performances, measure network resources, and record measurement data. The element manager reports the measurement-related information at scheduled frequency. The measurement result can be transferred among various network element management entities and stored locally.

The standardized northbound interface (Itf-N) between the element manager (or NEs with an embedded EM) and network manager is fundamental for a PLMN provider to manage multiple vendor network elements. A well-managed operator's network typically has its own network manager system and centralized network operation centers (NoC) that integrated the element managers from various component manufacturers, as well as from transport network providers.

Three essential components in this management model are performance management, fault management, and configuration management.

3.4.1 Performance Management

The performance management concerns the management of performance measurements and the collection of performance measurement data across a LTE network. It consists of administration of measurement scheduled by the network element manager, generation of measurement results in the network elements, and the transfer of these results to one or more operations systems.

The purpose of performance management activity is to collect data to verify the physical and logical configurations of the network and to locate potential problems as early as possible. Data is required to be produced by the NEs to support the following areas of performance evaluation:

- Traffic levels within the network, including the levels of user traffics and the signaling traffics; e.g., downlink and uplink data volume;
- Verification of the network configuration; e.g., transmit power levels;
- Resource access measurements; e.g., physical RB utilization and PDCCH utilization;
- Quality of Service; e.g., delays during call setup and packet throughput; and
- Resource availability; e.g., the recording of begin and end-times of service unavailability.

By the evaluation of the results produced by the measurements, operators understand the nature of the results received from the network, assure the reliability and accuracy of the measurement results, and ensure comparable measurement results for the same measurements being performed in equipment from different vendors. Operator can identify the results in the management systems with respect to the measurement jobs by the EM and the measurement types and measured resources by the NM. Operator can also take into account that, in a set of n correlated measurements, any $(n - 1)$ out of the defined n measurements may be provided by the network.

Performances measurements may also be used to supervise operator-defined threshold values and generate alarms upon exceeding the thresholds.

3.4.2 *Fault Management*

Fault management referred to the detection of faults, the generation, collection and presentation of alarms, operational-state data, and test results across network elements.

The fault management is accomplished by means of several processes, including the following:

- Fault surveillance and detection in the NEs;
- Notification of alarms (including alarm cease) and operational-state changes;
- Retrieval of current alarm and alarm history from the NEs, real-time forwarding of event reports;
- Fault isolation and defense mechanisms in the NEs;
- Alarm filtering;
- Management of alarm severity levels;
- Alarm and operational-state data presentation and analysis at the operations system (OS);

– Retention of alarm and operational-state data in the NEs and the OS; and the management of tests.

These processes are located within the network element and network element management layers. They are primarily proactive, rather than reactive for fault reported through external customer feedback. Since most underlying network infrastructure has the self-healing capabilities, FM can improve service quality and customer's perception of service with lowering costs.

The occurrence of failures in a NE may cause a deterioration of this NEs function and service quality, leading to the complete unavailability of the respective NE in severe cases. Network operator can minimize the effect of this failure by detecting failures in the network as soon as they occur and alerting the operating personnel as fast as possible. Then follow proper steps to isolate the failures (autonomously or through operator intervention), i.e., switch off faulty units or limit the effect of the failure by reconfiguration of the faulted NE and adjacent NEs. Network operator can also determine the cause of the failure using diagnosis and test routines, trigger healing procedure such as adjusting transmit power levels, bringing additional equipment into operation, reconfiguring existing equipment, or by repairing/eliminating the cause of the failure.

3.4.3 Configuration Management

Configuration management (CM) provides the operator with the ability to assure correct and effective operation of a PLMN network. CM controls and monitors the actual configuration on the network elements that may be initiated by the operator or by functions in the operations systems (OSs) or NEs.

Configuration management consists of both single network elements at NE management level and entire network configuration at network management level. It is essential that the configuration management will enable the operation of the multivendor network effectively and efficiently to support mix and match of multiple vendors' equipment.

CM actions may be requested as part of an implementation programme (e.g., additions and deletions), as part of an optimization programme (e.g., modifications), or simply configuration auditing to ensure the operational parameters are consistent with golden configuration in a commercial service network.

E-UTRAN management model consists of performance management, fault management, and configuration management. Both performance management and fault management are proactive and can be enhanced with analytics to further optimize network operation, while configuration management ensures the consistency among those operations. Standard Itf–N interface facilitates centralized performance management and fault management and self-healing systems across various network elements from different vendors.

3.5 RAN Network Key Performance Indicators and Analyses

Wireless network key performance indicators (KPI) are essential to ensure network normal operation and user experience. Various third parties actively survey and frequently measure throughput and latency among various commercial networks. Being able to claim the best and highest throughput network has been a vital competitive edge for an operator. This includes numerous Web sites for user speed and latency tests. Table 3.1 illustrates expected LTE performance.

User data throughputs depends on radio environment, cell loading, and device form factors. Figure 3.12 demonstrates the user throughputs of a loaded cell could be one-third of that of an unloaded network [reference]. It also showed that a properly engineered device could double the throughput of a poorly designed device.

Table 3.1 LTE specification in 3GPP

	3GPP specification
Data rates	Peak rate over 2 × 20 MHz DL: 150.8 Mbps (2 × 2); 299.6 Mbps (4 × 4); UL: 75.4 Mbps Peak rate over 2 × 100 MHz, 5 carrier aggregation: DL: 2998.6 Mbps (8 × 8); UL: 1497.8 Mbps
Spectral efficiency	DL 1.59 bps/Hz/sector UL 0.66 bps/Hz/sector
VoIP capacity	200+ voice users over 2 × 10 MHz
Intra-LTE handover	Interruption less than 20 ms
Call setup time	Less than 100 ms
Radio link latency	Less than 10 ms

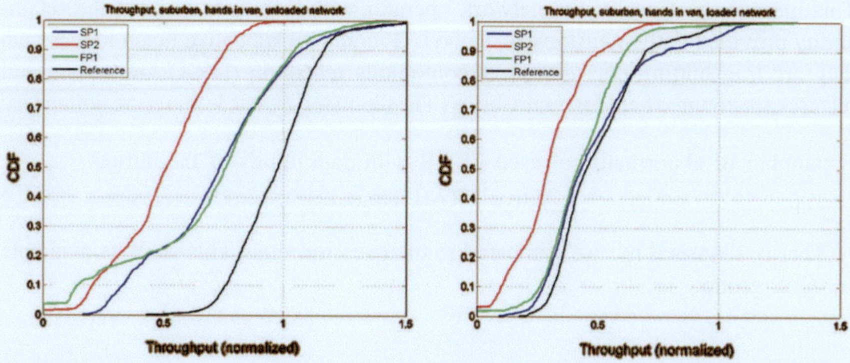

Fig. 3.12 E-UTRAN user performance. *Note* SP1/SP2 are two different smartphones, FP1 is a feature phone

Those user data rates are directly linked to a wide range of network key performance indicators and detailed measurement data operator monitoring in millisecond level. The essential key performance indicators in a mobile wireless network are accessibility, retainability, integrity, availability, and mobility.

Various services with different performance requirements also demand RAN network KPIs measured at each service level. This will require the KPI counters specified for each PLMN-ID, QoS, and SPID.

3.5.1 *Accessibility*

Accessibility represents the capability of network to support user radio access. It is also referred to as ineffective attempt or access failure rate. This measurement assists the network operator with information about the accessibility provided to customers. It is measured as a probability of success rate for E-RABs (E-UTRAN radio access bearers) establishment, that is, successful attempts compared with the total number of attempts for different parts of the E-RAB establishment:

$$\frac{\text{Number of Successful ERAB establishments}}{\text{Number of Received ERAB Establishment Attempts}} \tag{3.1}$$

Service can be mapped to initial or added E-RABs. It will also be measured for a specific QoS group, e.g., voice service and data service.

3.5.2 *Retainability*

Retainability shows how often an end user abnormally loses an E-RAB during the time the E-RAB is used. It is also referred as network drop call rate or lose call rate. This measurement assists the network operator with information about the retainability provided to their customers. Since LTE could be kept alive much longer than they are transmitting data due to discontinuous reception (DRX), only abnormal releases are counted here for retainability. The retainability rate is therefore defined as:

$$\frac{\text{number of abnormally released ERAB with data in any of the buffers}}{\text{Active ERAB Time}} \tag{3.2}$$

That is abnormal release compared to total session time. This shall be available per QoS group.

3.5.3 Integrity

Integrity represents user service quality during use of a service. For data service, this is typically represented by user throughputs and latency. For voice service, it can also be measured by equivalent Mean-Opinion Score and jitter. This measurement assists the network operator with information about the integrity provided to customers.

The service provided by E-UTRAN for this KPI shall be delivery of IP packets at IP layer to reflect user experience. It should also exclude the idle time to be independent of busty traffic pattern.

As illustrated in Fig. 3.13, the DL throughput can be calculated with total DL transmitted data volume divided by the time those data are transmitted,

$$\text{DL Throughput} = \frac{\sum_{\text{samples}} \text{Throughput Volume DL}}{\sum_{\text{samples}} \text{Througput Time DL}} \tag{3.3}$$

While the UL throughput can be calculated as total UL volume divided by UL transmission time:

$$\text{UL Throughput} = \frac{\sum_{\text{samples}} \text{Throughput Volume UL}}{\sum_{\text{samples}} \text{Throuput Time UL}} \tag{3.4}$$

Figure 3.14 shows the DL throughput as a function of signal-to-noise (SNR) ratio [reference]. As SNR increases, user-experienced throughput and sector throughput increase. The major objective of network optimization is to increase overall sector SNR across entire service area.

Those user throughput measurements also shall be available per QoS group. For voice service, additional integrity of latency could be measured by the time between reception of IP packet at the user device and the time when the eNB transmits the first block over the air interface:

$$\text{Latency DL} = \frac{\sum \text{Time\_Latency\_DL}}{\#\,\text{Samples}} \tag{3.5}$$

Voice quality shall also be measured by equivalent Mean-Opinion Score (eMOS), which is a calculation based on the performance of the LTE end-to-end network, e.g., defined in the ITU-T PESQ P.862 standard.

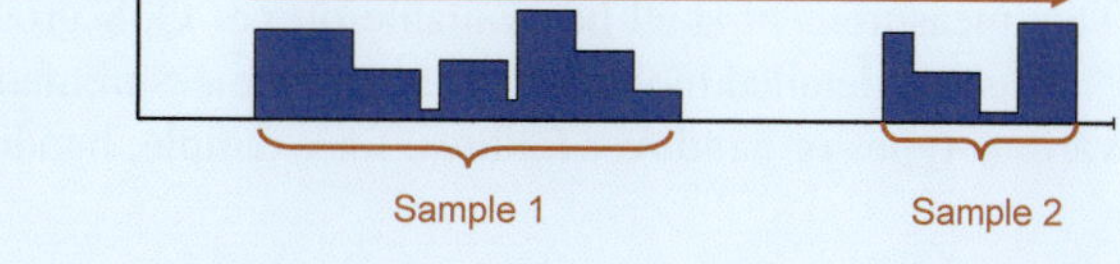

Fig. 3.13 Network performance measurement

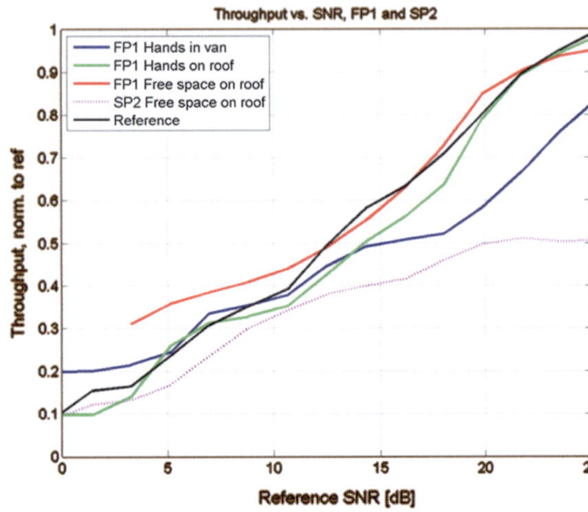

Fig. 3.14 Throughput versus SNR

3.5.4 *Availability*

Availability represents network robustness and reliability. It is also referred to network downtime. This measurement assists the operator with information about the availability of radio network. It is measured as the percentage of time that the cell is considered available:

$$\text{Availablity} = \frac{\text{Time that cell is available}}{\text{Measurement Time}} \times 100\% \tag{3.6}$$

This measurement reflects both coverage of a cell and operational well-being of a network.

3.5.5 *Mobility*

Mobility indicates the continuity of a wireless service. This measurement assists service operator on its mobility functionality and robustness:

$$\text{Mobility Success Rate} = \frac{\text{HO Execution Success}}{\text{HO Execution Attempt}} \times \frac{\text{HO Prep Success}}{\text{HO Prep Attempt}} \times 100\% \tag{3.7}$$

This measurement shall be available on per QoS group.

Various detailed mobility performance measurements are also available to define various types of handover failures, for example, handover failures due to handover to a sector too early, too late, and handover to a wrong cell. Handover interruption time on control and data plane can also being measured and reported.

On top of above KPIs to reflect the user perception and experience, various network performance status are also collected to enable high network efficiency and proper operation. Those could include the following:

- DRB and SRB, measurements related to data radio bearer and signaling radio bearer, respectively;
- RRC and RRU, measurements related to radio resource control and utilization, respectively;
- ERAB and S1SIG, measurements related to E-RAB and S1 signaling, respectively;
- HO, measurements related to handover;
- PAG, measurements related to paging;
- EQPT, measurements related to equipment;
- UECNTX, measurements related to User Equipment CONTEXT; and
- TB and MR, measurements related to Transport Block and Measurement Report, respectively.

Figure 3.15 illustrates a snapshot of downlink user throughput and drop call rate of four cells in an operational network. The drop call is typically higher at busy hour, 4–5 pm, while the user throughput is high when the cell is not fully loaded in early morning.

From those measurements, operator can determine the cell level capacity, control and data channel utilization, and traffic volume. Those are essential data points for

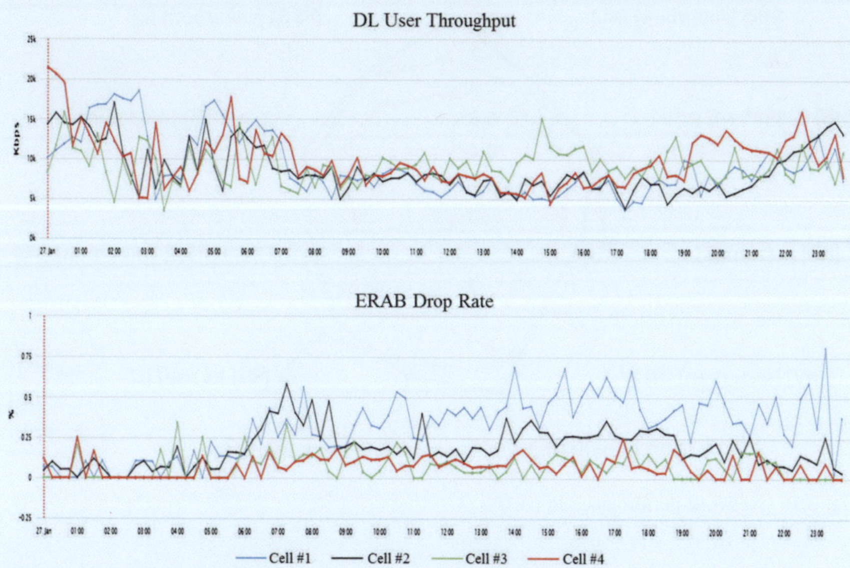

Fig. 3.15 Network performance measurement (examples only)

capital planning. Those measurement data can also be further analyzed to provide network and device design guideline to further improvement network performance.

Device plays an important role in a wireless network. The user throughputs of three difference devices are analyzed based on the network performance monitoring. Figure 3.16 shows the throughput normalized to reference antenna throughput (on same drive routes). Measurements for free space on roof (FSR), hands in van (HV), hands on roof (HR) use modes in combination with loaded (L) and unloaded (NL) network settings are included. Measurements using reference and all mockups are included. The Smartphone 1 (SP1) represented by blue curve demonstrated much better performance than that of Smartphone 2 (SP2) represented by red curve. A well-designed simple feature phase (FP1) could have similar performance as a well-designed smartphone. Figure 3.16 shows an urban sector with better SNR can support higher throughput than that of a larger inter-site distanced suburban cells.

Further analyses of various network performances can provide network dimensioning and intelligent control. Figure 3.17 shows an example of correlation of RRC setup failure % and average number of active connections in a cell. Those types of data could have capacity planning specific for a cell.

Operator can also obtain detailed network performance and cell trace to ensure proper daily operation, such as received signal strength, signal-to-noise ratio, and number of active users and radio connections, PDCCH and PUCCH utilization and error rates. A centralized self-organizing network can utilize the information to optimize network in real time.

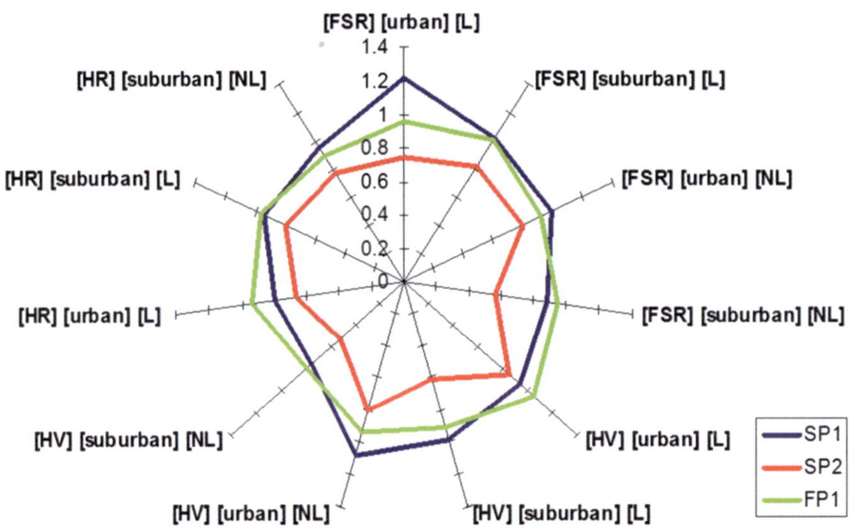

Fig. 3.16 Network performance impacts

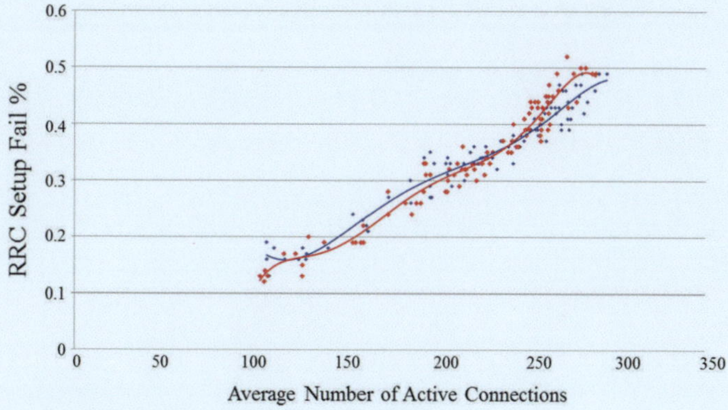

Fig. 3.17 Network performance impacts

3.6 Summary

E-UTRAN has standardized interfaces among all network elements. It has built architecture to support exponential traffic growth and mix-and-match of various network elements. The air interfaces have flexible and dynamic channel and resource management structures to enable various applications and services. Base station is more programmable and more adaptive to various environments and services. Those established foundations allow fine granularity and large scale of network performance and fault management model. Therefore, E-UTRAN provides much great opportunities for more advanced performance and fault analytic system to achieve optimized network performance and user experience.

Chapter 4
Enhanced Packet Core Network

Enhanced packet core (EPC) has evolved to support versatile of services. Separation of signaling and control plane, sophistical policy, and authentication ensure an efficient and secured core network. Legacy single EPC core is evolving to multiple EPC cores with software defined network (SDN) and network functions virtualization (NFV) capabilities to support IP connectivity, Internet, IMS, Internet of Things (IoT) with various human and machine communications services.

In this chapter, we will discuss EPC architecture in Sect. 4.1, followed by EPC interface specifications in Sect. 4.2, EPC network management model in Sect. 4.3, and then present EPC network key performance indicators and analyses in Sect. 4.4. Chapter summary is given in Sect. 4.5.

4.1 EPC Architecture

EPC architecture has been dramatically improved and expanded from voice centric to support of data, voice, multimedia, and machine-to-machine services. The architecture has the characteristics of flatness, manageability, and simplicity. LTE networks separate data and control planes, enable more efficient and flexible mobile wireless communications.

Figure 4.1 illustrates LTE EPC architecture. Evolved Node B has radio link control and IP tunneling interfaces. Radio link control path (S1-MME) toward mobility management entity is split from data path (S1-U) toward serving gateway. There is no physical layer beyond eNB. MME has control function for mobility management. User plane data is further transmitted from serving GW to packet data network gateway through S5 for intra-operator and S8 for inter-operators. Serving gateway and PDN-GW have interfaces with policy and charging rules function (PCRF) through Gxc and Gx, respectively. Gx provides transfer of quality of service policy and charging rules from PCRF to policy and charging enforcement

D. Kakadia et al., *Network Performance and Fault Analytics for LTE Wireless Service Providers*, DOI 10.1007/978-81-322-3721-1_4

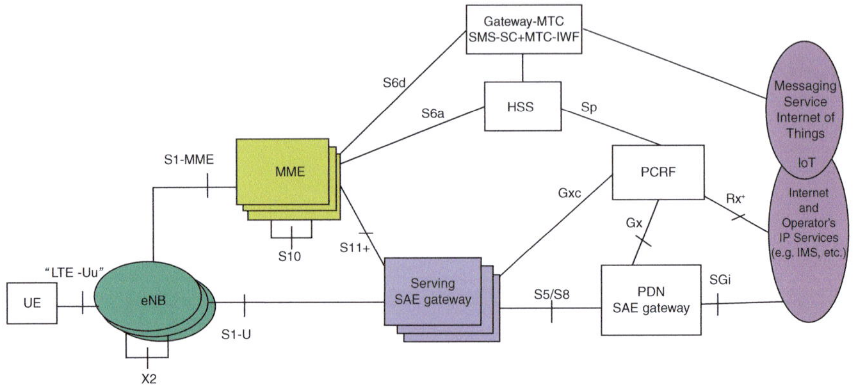

Fig. 4.1 LTE EPC architecture

function (PCEF) resided inside PDN-GW. MME interfaces with home subscriber server (HSS) through S6a.

The interfaces in Fig. 4.1 are configured as one-to-many fashion. One UE can be connected to many eNBs. One eNB can be connected to multiple MMEs and multiple SGWs. One MME is connected to multiple eNBs and SGWs. And one SGW is connected to many eNBs and MMEs. Standardized interfaces are also defined across different eNBs through X2, different MMEs through S10, and different SGW through S5 to allow mix-and-match of various network elements. Those more open and flat RAN interfaces enable various deployment scenarios, such as macro, micro, pico, and femto. It support independent scaling of control and user plane functions, for example, control plane data in MME could be more scaling with the number of users, while user plane SGW could be scaling with traffic volume. It also makes centralized control mechanism and performance management viable. S11 is the reference point between MME and Serving GW. SGi is the reference point between the PDN-GW and the packet data network. Packet data network may be an operator external public or private packet data network or an intra-operator packet data network, e.g., for provision of IMS services. The Rx reference point resides between the application function and the PCRF.

Internet service is the dominant service right now for smartphone. IP multimedia sub-system (IMS) is another example of services on top of the IP connectivity to support voice and video telephony applications.

EPC can support various messaging services. It can support legacy short message service (SMS) through SMS-service center (SMS-SC) with MME S6d interface. It can also expand the existing SMS-SC to support machine-type communication (MTC) through MTC inter-working function (MTC-IWF) or gateway function for MTC as specified in small data architecture in TS23.682.

Functionalities are diligently divided among various network elements in a LTE EPC network. We are going to describe in detail the SAE gateways, MME, PCRF, HSS, and application services.

4.1.1 SAE Gateways

System architecture evolution (SAE) gateways consist of serving gateway and PDN gateway. They are the mobile wireless access gateway to Internet and operator's IP services. They optimize packet flow with QoS markings, enable policy decision, and enforcement. They perform packet routing, user plane tunneling management, and data buffering and forwarding. They also take care of security management, lawful interception, authentication, charging, and accounting.

The two logical SGW and PDN gateways may be implemented in one physical node or separated physical nodes. When the S5/S8 interface is based on GTP, the SGW will only support GTP interface, while PGW is left to map IP service flows to GTP tunnels and to connect PCRF. When S5/S8 interface is based on PMIP, the SGW will perform mapping between IP service flows and GTP tunnels and will connect to PCRF for the mapping information.

Serving GW is the gateway which terminates the interface toward E-UTRAN. For each UE associated with an evolved packet system (EPS), there is only one serving GW at a given time. Serving GW supports packet routing and mobility anchoring for both GTP-based and PMIP-based S5/S8.

SGW packet routing and forwarding are based on the requests from MME, PGW, and PCRF. It marks transport-level packet in the uplink and downlink, e.g., setting the DiffServ code point, based on the QCI of the associated packet network bearer. It maps IP flows in S5/S8 interface to bearers in the S1-U interface for bearer binding and event reporting function (BBERF).

SGW is the local mobility anchor point for inter-eNodeB handover. The MME commands the SGW to switch the tunnel from one eNB to another, or even one SGW to another SGW. It sends one or more "end marker" to the source eNodeB immediately after switching the path during inter-eNodeB handover, especially to assist the reordering function in eNodeB. It also acts as mobility anchoring for inter-3GPP mobility, relaying the traffic between 2G/3G system and PDN-GW.

When a UE is in connected mode, the SGW only relays the data between eNB and PGW. When the UE is in idle mode, the resource in eNB is released and the data path terminates at the SGW. Therefore, if SGW receives data packets from PGW, it will buffer the packets and request the MME to initiate paging of the UE. The paging will lead the UE to reconnect. Once the UE connection is established, the buffered packets will be sent to the eNB.

The PDN-GW is the gateway which terminates the SGi interface toward the PDN. If a UE is accessing multiple PDNs, there may be more than one PDN-GW for that UE. PDN-GW functions include both the GTP-based and the PMIP-based S5/S8.

PGW can allocate the IP address to a UE, and the UE used this IP address to communicate with other IP hosts, such as Internet. An external PDN can also allocate an IP address to the UE it connected to, and then PGW will tunnel all traffic to that network. The PGW can perform the required dynamic host configuration

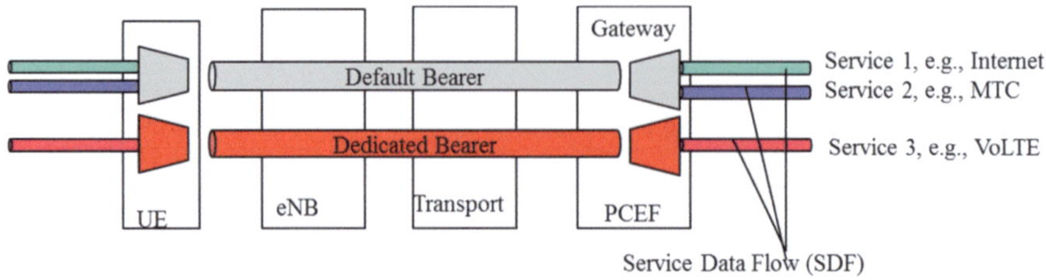

Fig. 4.2 LTE EPC architecture

protocol (DHCP) functionality, or query an external DHCP server, and deliver the address to the UE.

PGW performs policy and charging enforcement function (PCEF). It enables UL and DL gating and filtering function per user as required by the policies set for the UE and the service. It marks the transport-level packet in the uplink and downlink based on the QCI of the associated packet bearers. The PGW also supports rate enforcement for rate policing and shaping per service data flow (SDF) as shown in Fig. 4.2. The gray flow is a default non-GBR bearer, while the red flow is a dedicated GBR beater. PGW can enforce access point name (APN) aggregate maximum bit rate by rate policing/shaping per aggregate of traffic of all SDFs of the same APN that are associated with non-GBR QCIs. PGW can also enforce DL rate based on the accumulated access maximum bit rate (MBRs) of the aggregate of SDFs with the same GBR QCI.

PGW can perform lawful interception. It accounts for inter-operator and intra-operator charging. PGW collects and reports the uplink and downlink data volume (per EPS bearer) as received from and sent to the serving node. It supports UL and DL service-level charging based on SDFs defined by the PCRF, or based on deep packet inspection defined by local policy. It provides accounting per UE and bearer.

PGW is the highest level mobility anchor in the system. When a UE moves from one SGW to another, the bearers have to be switched in the PGW. PGW will send one or more “end marker” to the source SGW immediately after switching the path during SGW change. It also supports UL and DL bearer binding and UL bearer binding verification. The PGW provides PDN connectivity to both GERAN/UTRAN only UEs and E-UTRAN capable UEs using any of E-UTRAN, GERAN, or UTRAN.

4.1.2 Mobility Management Entity

Mobility management entity controls eNB, MME, SGW, and PGW selection and mobility. The mobility includes cell handover through S1 and X2 interfaces, inter-technology handover, and idle-mode handover. MME manages data bearer establishment, session setup, network access control, and authentication. It is responsible for user tracking and paging, NAS signaling, and signaling security.

Mobility management is the major function of MME. The MME tracks the locations of all UEs in its service area. MME will create an entry for the UE when a UE is registered in the network and signal the location to the HSS in the UE home network; then, it requests proper resource setup at its selected eNB, SGW, and PGW. The MME will keep tracking the UE location at eNB level when the UE is in connected mode. The MME will keep tracking the UE at the level of tracking area (TA)—a group of eNBs, when the UE goes to idle mode. An idle-mode UE will report its location either periodically or when it moves to another tracking area. When data is received for an idle UE, the MME will be notified, and MME will page the UE in the TA to initiate communications. MME manages the tracking area list. MME actively involved in control signaling and every steps of handover of an active-mode UE between eNBs, SGWs, and MMEs. It supports relaying function including relay nodes attachment and detachment.

Session and subscriber management are essential parts of MME. MME initiates the authentication with the UE home network home subscription server (HSS) when a UE registers to the network. MME is in charge of NAS signaling and NAS signaling security. MME retrieves subscription profile from the UE home network when UE is registered. MME will store this information and use this profile to determine what type of service quality and packet network connection should be allocated to the UE. It supports bearer management functions, including dedicated bearer establishment and setup roaming S6a toward home HSS. It supports lawful interception of signaling traffic. It supports change of UE presence in presence reporting area reporting.

MME can also support short data service and avoid data bearer establishment for a short data message. This is particularly useful for machine-type communications or message services dominated with less than 250 bytes of data. SMS over NAS signaling is sent together with RRC connection completion message. This can save total signaling message by half. However, the message would be sent immediately with no scheduling gain or multiple-stream transmission.

The serving GW and the MME may be implemented in one physical node or separated physical nodes. MME typically manages thousands of eNBs, located in a more centralized location.

4.1.3 Policy and Charging Resource Function (PCRF)

PCRF dynamically controls and manages all data sessions and provides appropriate interfaces toward charging and billing systems, as well as enables new service revenues. It is the policy and charging control element that is responsible for policy and charging control (PCC). It determines how to handle the service in terms of QoS and pass the information to the PCEF located inside PGW, so that proper bearer and policy can be set up.

In non-roaming scenario, there is only one single PCRF associated with one UE IP session. The PCRF terminates the Rx interface and the Gx interface. In a roaming scenario with local breakout of traffic, there may be two PCRFs associated with one UE IP session: H-PCRF that resides within the home network and V-PCRF that resides within the visited network.

The functions of the home PCRF (H-PCRF) described in TS23.203 include terminating the Rx reference point for home network services, terminating the S9 reference point for roaming with local breakout and associating the sessions established over the multiple reference points (S9, Rx), for the same UE IP session (PCC session binding). While in a visited PCRF (V-PCRF), it only involved terminating the Gx and S9 reference points for roaming with local breakout and terminating Rx for roaming with local breakout and visited operator's application function.

Policy control is increasingly important as the operator needs to define service priority and ensure proper service performance for various services. This is in particular critical as we are introducing billions of low-latency, low-priority machine-type devices with very different characteristics as traditional voice and data services.

Policy control and charging architecture is shown in Fig. 4.3, where PCRF is in the center. PCC enables a service session (also called IP connectivity access network or IP-CAN) with appropriated bandwidth and QoS. It also provides a means to control charging on a per-service basis. The PCRF provides network control regarding service data flow detection, gating (blocking or allowing packets), QoS control, and flow-based charging toward the PCEF resided at PGW. It can also

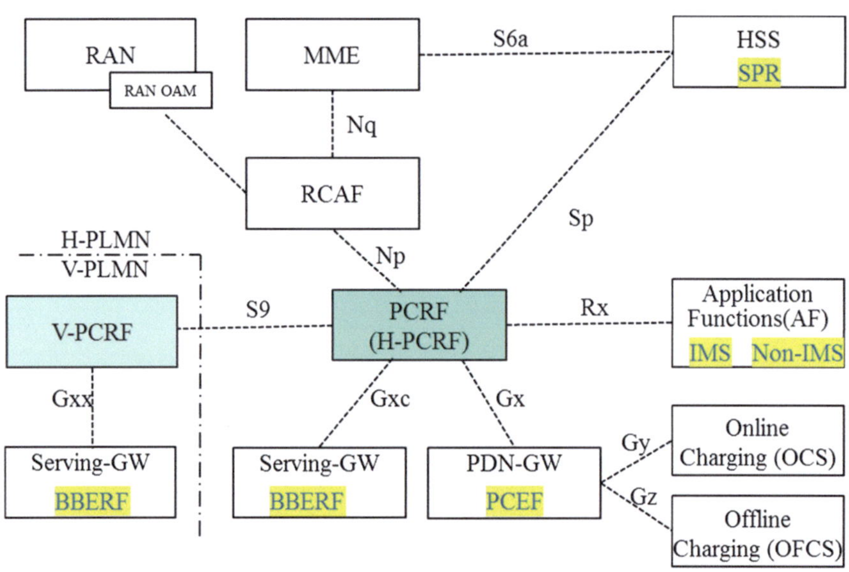

Fig. 4.3 LTE PCC architecture

apply security procedures before accepting information from the application function (AF). The PCRF ensures that the PCEF user plane traffic mapping and treatment is in accordance with the user's subscription profile which it receives from subscriber profile repository (SPR) resided at HSS over Sp interface. The PCRF may reject the request received from the AF when the service information is not consistent with subscription information (either locally configured or received from SPR) and the PCRF responds to the AF with appropriate reason.

RAN congestion awareness function (RCAF) is introduced in 3GPP Rel. 13 to mitigate user plane congestion. The RCAF collects per cell measurements from RAN OAM, retrieves list of UEs per cell/eNB and list of active APNs per UE from MME via Nq interface. The RCAF then determines serving PCRF(s) based on IMSI/APN and reports congestion via Np interface. The PRCF will adapt policies (via Gx/Rx) interface to activate application optimization mechanism, such as video compression, and to defer services in case of RAN congestion.

4.1.4 Home Subscription Server (HSS)

Home subscription server is the subscription data repository for all permanent user data. HSS manages subscription-related information in real time for multiaccess and multidomains in an all-IP environment. This includes managing subscriber identities, service profiles, APN entitlements, and QoS profiles for access to the LTE network and IMS applications.

HSS supports the network control layer with subscription and session handling, providing capabilities for mobile management, use security, use identification handling, access authorization, and service authorization in both evolved packet core and generic access network (GAN) networks. The subscriber information contains the services that are applicable to the user, the allowed PDN connections and services, permission to roaming into a particular visited network. The permanent key used to calculate the authentication vectors for user authentication and deriving subsequent keys for encryption and integrity protection is stored in the authentication center typically inside HSS.

HSS communicates with PCRF to determine policy and subscriber information. HSS directs MME to execute the control information through S6a interface.

Most advanced HSS is modular and can be adapted to manage multiaccess scenarios, providing optimal integration of packet-switched technologies. It can be deployed locally or virtualized.

4.1.5 Application Services Domain (AS)

Application services (AS) domains include various sub-systems for application services. Application server is a piece of middleware that connects back-end systems to mobile devices. It can be Web server, Internet server, security server, and voice, video, and machine-type service servers. There are two types of application domains:

1. The IP multimedia sub-system (IMS) is service machinery that operator uses to provide services using the session initiation protocol (SIP). IMS has 3GPP-defined architecture to support voice over LTE (VoLTE), and video and multimedia services
2. Non-IMS referred to traditional data services as well as emerging Internet of Things (IoT). The data services could be messaging SMS/MMS, IP messaging, Internet, and Web 2.0, as well as position/location services and mobile TV. Internet of Things (IoT) server provides services for machine-type applications. For example, it can be connected with location server to provide device tracking and location logs.

On top of data, voice and video services, the EPS has been expanded to support emergency service, priority service, home eNB functions (or femto), local IP offload, messaging service, WLAN service, and machine-type services.

Both IMS and non-IMS services consist of communication services and messaging services. They require dynamic policy management and real-time service control. Flexibility and scalability of application servers enables rapid introduction of new services to enable new service revenues.

4.2 EPC Interfaces Specifications

EPC network has achieved the network efficiency and manageability through well-defined fully standardized interfaces. This allows flexibility deployment of various nodes and mix-and-match network elements from multiple manufacturers.

4.2.1 S1-U Interface

S1-U interface is the user plane interface between eNB and SGW. S1-U interface followed GPRS tunneling protocol for the user plane (GTP-U). MME controls the user plane tunnel establishment and established user plane bearers. GTP shall encapsulate all end user IP packets.

UDP/IP is the backbone network protocol used for routing user data and control signaling. User datagram protocol (UDP) transfers user data, defined in RFC 768. LTE-Uu is radio protocol of E-UTRAN between the UE and the eNodeB specified in TS36.300.

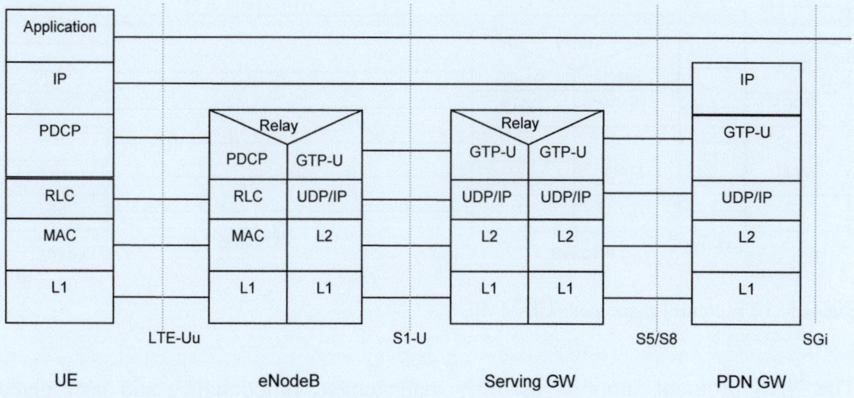

Fig. 4.4 LTE user plane path UE-PGW

User plane traffic is carried over virtual connections through service data flow. SDFs are carried over bearers with unique QoS characteristics as described in session 4.1.3. One bearer has three segments as illustrated in Fig. 4.4:

- Radio bearer between UE and eNB
- Data bearer between eNB and SG over S1-U interface
- Data bearer between SGW and PG over S5 interface.

4.2.2 S1-C Interface

S1-C interface is the control plane interface between eNB and MME. The control plane consists of protocols for control and support of the user plane functions:

- Controlling the E-UTRA network access connections, such as attaching to and detaching from E-UTRAN;
- Controlling the attributes of an established network access connection, such as activation of an IP address;
- Controlling the routing path of an established network connection in order to support user mobility;
- Controlling the assignment of network resources to meet changing user demands.

Both eNB-MME and UE-MME control planes are used in E-UTRAN mode. S1 application protocol (S1-AP) defines the application layer protocol between the eNodeB and the MME. It also uses stream control transmission protocol (SCTP) to guarantee delivery of signaling messages between MME and eNodeB. SCTP is defined in RFC 4960. The highest layer in S1-MME interface is NAS layer.

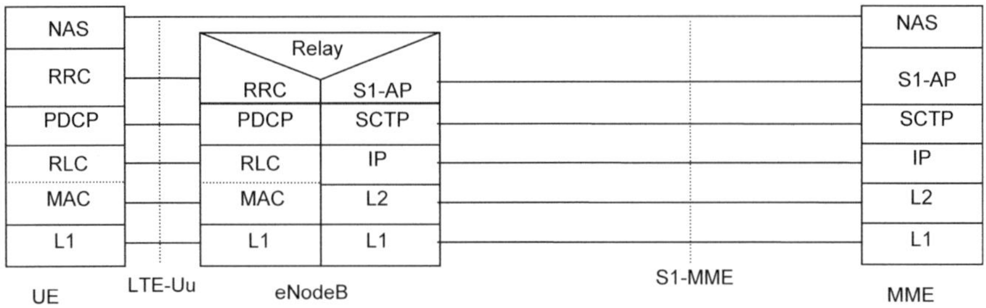

Fig. 4.5 LTE control plane path UE-MME

The NAS protocol supports mobility management functionality and user plane bearer activation, modification, and deactivation. It is also responsible for ciphering and integrity protection of NAS signaling (Figure 4.5).

4.2.3 S5/S8 and SGi Interfaces

S5 and S8 interfaces are the user plane interfaces between SGW and PGW. The S5 refers to the user not roaming, i.e., user is attached to the home network. The S8 refers to the user is roaming, or user is attached to a visited LTE network.

There are two different protocol variants for S5/S8 interfaces: one is GTP protocol,the same protocol used to provide IP connectivity over GSM/GPRS and WCDMA/HSPA networks; and the other is IETF PMIPv6 protocol. The dominant de facto protocol used in commercial deployment is GTP protocol.

Operator may choose to deploy equipment that can combine SGW and PGW functionalities whenever needed. This will eliminate the S5 interface and reduce the amount of hardware to process the user data plane by up to 50% in theory.

In other user cases, S5 is essential to split gateway deployment. This could be due to the user's need to connect to more than one external data network at the same time, and not all of these can be served from the same PGW. All user data will then connect to the same SGW, but more than one PGW.

S5 interface also support more flexibility deployment of more distributed SGWs and centralized PGW, so that SGW is more close to eNB to reduce latency and backhaul cost. PGW also can serve as an anchor point, when a user is moving between two LTE radio base stations that does not belong to the same service area and connected to two different SGWs.

SGi interface is the interface between PGW and external IP networks. The SGi interface supports various services. The protocol stack on SGi is the application protocol over IP. For example, the protocol stacks could vary to support normal unicast service, MBMS service, and position server.

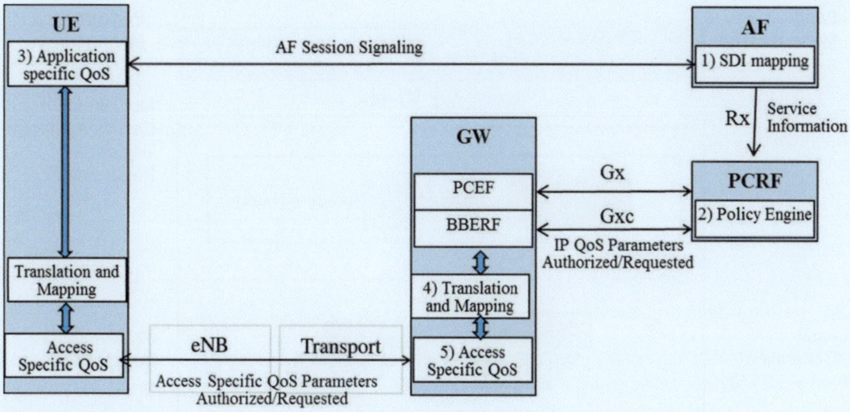

Fig. 4.6 QoS mapping for UE-initiated IP-CAN bearers

4.2.4 Gx/Gxc and Rx Interfaces

Gx and Gxc play an important role in policy, charging management, and QoS service establishment.

Here, we use a UE-initiated bearer establishment as an example to show interworking among various interfaces. When a UE desires to establish or modify a bearer, the following steps are followed as illustrated in Fig. 4.6.

1. The application function (AF) maps session description information (SDI) within the AF session signaling to service information and passes to the PCRF over the Rx interface.
2. The PCRF shall map from the service information received over the Rx interface to the authorized IP QoS parameters that is passed to the PCEF via the Gx to PGW, and BBERF via Gxc interface to SGW. The mapping is performed for each IP flow. Upon a request from the PCEF/BBERF, the PCRF combines the individual-authorized IP QoS parameters per flow.
3. The UE derives access-specific QoS parameters from the AF session signaling in an application-specific manner. The IP and access-specific QoS parameters are generated according to the application demands.
4. The PCEF/BBERF shall map from the authorized IP QoS parameters received from PCRF to the authorized access-specific QoS parameters.
5. The PCEF/BBERF shall compare the requested access-specific QoS parameters against the authorized access-specific QoS parameters.

The mapping that takes place in the UE and the network should be compatible in order to ensure that the PCEF will be able to correctly authorize the session. Figure 4.6 shows different kinds of QoS parameters in different points of QoS mapping. Due to the UE requests, there are bidirectional flows between the UE and the PCRF.

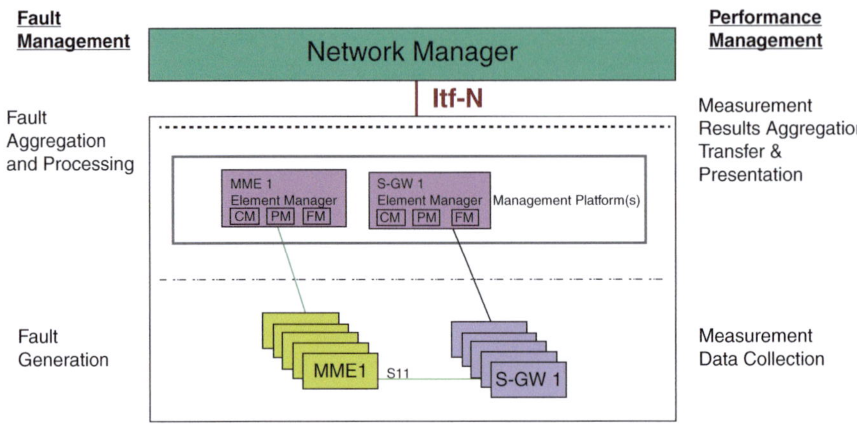

Fig. 4.7 EPC network management Model

4.3 EPC Network Management Model

EPC network management follows the same model as eNB in session 2.4 with performance management, fault management, and configuration management associated with EPC equipment, mobility, IP, inter-RAT, QoS, security, session, and subscriber measurements. The network management shall provide network performance for every PLMN-ID and every QoS or SPID under each PLMN-ID. Typical operator network management network sniff into all the interfaces in Fig. 4.1 to provide performance measurement and diagnostics (Figure 4.7).

4.3.1 MME Mobility Measurements

MME mobility measurements include EPS attach procedure, UE-/MME-/HSS-initiated EPS detach procedures, tracking area update with and without serving GW change, paging, overloading, EMM-registered subscriber numbers, handover, routing area update, combined tracking area/location area update and number of implicit detach initiated by MME.

EPS attach procedure can be measured as attempted EPS attach, successful EPS attach, and the failed attach procedures. The failed EPS attach procedures are distinguished by reject cause. Possible reject causes are defined in TS24.301. The EPS attach procedure also includes combined EPS/IMSI attach and emergency attach.

EPS detach procedures include the following three categories:

- UE-initiated EPS detach procedure reports attempted and successful EPS detach procedures by UE. They provide the number of attempted and successful EPS detach procedures initiated by UE.
- MME-initiated EPS detach procedure reports attempted and successful EPS detach procedures by MME.
- HSS-initiated EPS detach procedure provides attempted and successful EPS detach procedures initiated by HSS.

Tracking area update procedure with serving GW change reports attempted, successful and failed tracking area updated procedures with serving GW change. The failed tracking area update procedures will response to "Tracking Area Update Request" with a "TRACKING AREA UPDATE REJECT" message. In this reject message, it will include the last visited registered routing area identification (RAI) information element indicated to the MME that it wishes to be served by a new serving GW.

Tracking area update procedure without serving GW change reports attempted, successful, and failed tracking area updated procedures without serving GW change. The failed tracking area update procedures will response to "Tracking Area Update Request" with a "TRACKING AREA UPDATE REJECT," including it wishes to be served by the same serving GW as the old serving GW with corresponding incremental reject cause.

EPS paging procedures reports attempted, successful, and failed EPS paging procedures. The number of successful and failed EPS paging procedures initiated at the MME is measured. The initial paging procedures as well as the repeated paging procedures are counted.

MME control of overload-related measurements for EPC reports attempted overload start procedures and stop procedures. It measured transmission of "Overload Start" and "Overload Stop" message from MME to each eNB according to TS23.401 and TS36.413, respectively.

EMM-Registered subscribers also reported by MME includes the mean and maximum number of EMM-Registered subscribers are obtained by sampling at a pre-defined interval the number of EMM-Registered subscribers in a MME and then taking the arithmetic mean and maximum, respectively.

Handover-related measurements include inter-RAT handovers among various technologies, including LTE, GSM, UMTS, CDMA2000, and combination.

Routing area update with MME interaction reports the number of attempted routing area update with MME interaction initiated within this MME area. It also measures the successful routing area update with MME interaction with and without SGW change, respectively, as well as the failed routing area update with MME interaction with and without SGW change, respectively. The failed measurements are split into sub-counters per the reject cause. The rejection cause is defined in TS29.274.

Combined TA/LA update procedure reports an attempted, successful, and failed combined tracking area/location area update. This measurement is triggered when MME receives "Tracking Area Update Requisition" from a UE indicating this is a combined TA/LA update request or a combined TA/LA update with IMSI attach request, TS23.272. The failed TA/LA update will be split into sub-counters per the reject cause.

The number of implicit detach-related measurement initiated by MME is reported by MME. MME initiates "Implicit detach" in order to return the EMM context to EMM-deregistered state (TS24.301).

4.3.2 MME Session and Subscriber Management

MME session management reports the number of dedicated EPS bearers in active mode, dedicated bearer setup time, MME-initiated dedicated bearer activation and modification, total EPS service request, MME subscriber number in idle mode and connected mode, number of incoming and outgoing IP packets, location update, mean and peak usage of key processors.

The number of dedicated EPS bearers in active-mode mean and maximum is reported by sampling at a pre-defined interval the number of dedicated EPS bearer established by MME in active mode and then taking the arithmetic mean and maximum, respectively.

Dedicated bearer setup time by MME is measured by accumulating the time intervals for every successful dedicated bearer setup by MME between the MME transmission of a "Activate dedicated EPS bearer context request" and the corresponding "activate dedicated EPS bearer context accept" message receipt by the MME. The end value of this time will then be divided by the number of successful dedicated bearer setup procedures to give the arithmetic mean.

MME-initiated dedicated bearer activation includes attempted, successful, and failed dedicated bearer activation procedures by MME. The failed cases will be incremented according to the reject cause TS24.301.

MME-initiated EPS bearer modification included attempted, successful, and failed EPS bearer modification procedures by MME. The failed cases will be incremented according to the reject cause. MME will send UE "Modify EPS bearer context reject" message with "ESM Cause" taking the reject cause.

Total EPS service requests include both UE-initiated and network-initiated service requests. It consists of total attempted, successful, and failed EPS serve request procedures, respectively. MME will send "Service Reject" message to eNB as a result of Service Request procedure with possible reject causes defined in TS24.301.

MME subscriber management consists of insert subscriber data requests received from a HSS. MME will report the number of attempted insert, delete subscriber data requests received from a HSS, respectively. It will report the number of subscribers in ECM-idle state. This measurement is split into sub-counter per track area (TA). If a TA list contains multi-TA, this sub-counter can be measured in current TA in which subscribers were in ECM-connected state last time. It will also report the number of subscribers in ECM-connected state in sub-counters per track area.

MME also reports the number of incoming and outgoing IP data packets on the MME S10 interface from eNB to MME, and MME to eNB, respectively. It will also measure the number of octets of incoming and outgoing IP data packets on the S1-MME interface from eNB to MME, MME to eNB, respectively. In this case, the data packet size is extracted from the IP header and added on to the measurement value, see TS36.412.

MME will update location-related measurements including attempted, successful, and failed update location procedure. The failed update location procedure will be received by MME on S6a interface from HSS to indicate failure cause.

MME shall also report the mean and peak usage of each key processor during the granularity period. A MME may have more than one key processor. The measurement will sample at a pre-defined interval the usage of the processor and then taking the arithmetic mean and maximum for each key processor.

4.3.3 SGW Measurements

SGW measurements include SGW-GTP-related measurement, S1-U-related measurement, and SGW session measurement.

GTP S5/S8-related measurement includes the number of outgoing GTP data packets on the S5/S8 interface from SGW to PDN-GW, the number of incoming GTP data packet on the S5/S8 interface from PDN-GW to SGW, as well as the number of octets of outgoing GTP data packets on the S5/S8 interface, from SGW to PDN-GW and incoming packets from PDN-GW to SGW. The GTP signaling packets are also reported on both incoming and outgoing between PDN-GW and SGW on both data and octets.

S1-U data volume-related measurements include the number of outgoing GTP data packets on the S1-U interface from SGW to eNB. This provides the number of GTP data PDUs on the S1-U interface which have been generated by the GTP-U protocol entity on the S1-U interface. The corresponding incoming packets are also measured in the direction from eNB to SGW. The number of octets of corresponding packets is measured.

SGW session measurement monitors EPS bearer creation. EPS bearer creation-related measurements on S11 interface are generated to provide the attempted, successful numbers of default bearer creation, dedicated bearer creation, and EPS bearer modification. It also measures the S5/S8-related default bearer creation, dedicated bearer creation, bearer medication, and bearer deletion. Bearer resource usages are provided in terms of mean and max number of active EPS bearers.

4.3.4 PDN-GW Management

PDN-GW measurement includes bearer creation, session management, and SGi-related measurement.

PDN-GW-initiated dedicated bearer creation includes the attempted, successful, and failed number of PDN-GW-initiated dedicated bearer creation. The failed PDN-GW-initiated dedicated bearer creation is split into sub-counters as per the reject causes specified in TS 29.274 and TS23.401. The same is also recorded for attempted, successful, and failed number of PDN-GW-initiated dedicated bearer deletion, PDN-GW-initiated dedicated bearer modification with and without QoS update.

Session management measures the mean and max number of active EPS bearers, UE-requested bearer resource modification procedure, mean and max number of PDN connections per APN. It also recorded the mean and max number of EPS bearers, split into sub-counters of both dedicated bearer and default bearer.

SGi-related measurement includes SGi incoming and outgoing link usage to provide IP-layer incoming and outgoing link usage of SGi interface, respectively.

4.3.5 PCRF-Related Measurement

PCRF-related measurement consists of IP-CAN session establishment, session modification, authentication of QoS, Gateway control session establishment, credit re-authorization procedure, and IP-CAN session termination.

IP connectivity access network session establishment-related measurements include attempted, successful, and failed IP-CAN session establishment. The failed IP-CAN session establishment is split into sub-counters as per the failure cause. Possible failure causes are included in RFC 3588.

IP session modification-related measurements include attempted, successful, and failed IP session modification. When PCRF transmits a Credit-Control-Answer (CCA) Diameter command to PCEF with non-2xxx result code in the UE-initiated IP session modification procedure (PCC Rule Provisioning in PULL Mode), each failed IP session medication will be added to the relevant measurement according to the failure cause.

Authorization of QoS-related measurements is quite essential for PCRF. The attempted, successful, and failed resource authorization procedures at session establishment and session modification are recorded. Each failed resource authorization procedure is added to the relevant measurement according to the failure cause.

Gateway control session establishment-related measurements include attempted, successful, and failed gateway control session establishments, while failed control session establishment will be sub-countered according to failure cause.

Credit re-authorization procedure-related measurement includes attempted, successful, and failed credit re-authorization procedure. On receipt by the PCRF of a Re-Auth-Answer (RAA) Diameter command with result code $\neq$ 2001, from PCEF in response to the RAR command, each failed credit re-authorization procedure will be added to the relevant measurement according to the failure cause. Possible failure causes are included in RFC 3588. The sum of all supported per-cause measurements shall equal the total number of failed credit re-authorization procedures. In case only a subset of per-cause measurements is supported, a sum sub-counter will be provided first. See TS29.212.

IP connectivity access network session termination-related measurements include the attempted, successful IP session termination on the Gx interface.

4.4 EPC Key Performance Indicators and Analyses

Wireless network performance indicators are essential to ensure network normal operation and user experience. On top of the KPIs monitored at eNB in session 2.5, MME, SGW, and PGW are required to monitor end-to-end performance and network element capacity and status.

4.4.1 MME Accessibility, Mobility, and Capacity

The major function of MME is mobility management and session continuity. Thus, it is essential to ensure mobility and session performance. Table 4.1 illustrates major KPIs associated with accessibility and mobility.

MME capacity is defined by session capacity, including the number of sessions and throughput supported. Table 4.2 illustrates major MME capacity-related KPIs.

4.4.2 SGW and PGW Capacity and Connectivity

SGW and PGW performance is measured by their capacity, accessibility, and mobility. They are defined in Table 4.3.

From those measurements, operator can determine MME and SGW capacity, paging, TAU, and handover performance, and those are critical data points for capital planning and routing optimization.

Table 4.1 Major MME accessibility and mobility KPIs

	MME KPIs
Attachment	Attach success rate; attach failure rate (total) Attach failure rate due to system-related reasons
Paging	Paging success rate
Service request	Service request setup success rate Service request failure rate (total) Service request failure rate due to system-related reasons
Tracking	Tracking area update success rate (total) Intra-MME tracking area update success rate Inter-MME tracking area update success rate
Bearer setup	Default bearer setup success rate Defecated bearer setup success rate Update bearer success rate
Authentication	HSS authentication success rate UE authentication success rate
Mobility	Path switch success rate at MME at the same SGW Path switch success rate at MME at new SGW Incoming S1 handover success rate without SGW relocation at target MME Outgoing S1 handover success rate without SGW relocation at source MME S1 handover success rate with relocation at same MME S1 handover success rate without SGW relocation at same MME

Table 4.2 Major MME capacity KPIs

	MME KPIs
Connected UEs	Average number of connected UEs at MME Maximum number of connected UEs at MME
Registered UEs	Average number of registered UEs at MME Maximum number of registered UEs at MME
Defaults bearers	Average number of default bearers at MME Maximum number of default bearers at MME
Dedicated bearers	Average number of dedicated bearers at MME Maximum number of dedicated bearers at MME

4.4.3 Service-Specific Performance and Analyses

As network supports more and more varieties of applications, more service-specific end-to-end performance of those applications needs to be measured and analyzed. Table 4.4 lists major KPIs for service qualities of voice over LTE (VoLTE).

Those VoLTE KPIs are monitored and analyzed throughout commercial network operation. Figure 4.8 illustrates an example of voice quality during LTE handover periods. A VoLTE user crosses several sector boundaries with physical cell identification (PCI), ranging from 300 to 375, 125,150,160, etc. The chart demonstrated that voice quality stayed around 4.5 to 4 MOS majority of time even during handover period with higher packet lost. The jitter is around 20 ms without handover and around 30 ms during handover. High jitter led to low MOS.

Table 4.3 Major SGW and PGW KPIs

	SGW and PDN-GW KPIs
Capacity	#Sessions Throughput #UE #eNB #MME
Accessibility	Attach success rate Service request success rate Connection attempt success rate Paging success rate Network-initiated dedicated bearer activation success rate HSS-initiated QoS modification success rate
Mobility	Intra-SGW handover success rate Inter-SGW handover success rate via X2 Inter-SGW with indirect tunnel handover success rate Inter-SGW without indirect tunnel handover success rate Intra-SGW HO success rate TAU Inter-SGW handover success rate TAU eHRPD-to-LTE handover success rate

Table 4.4 Major VoLTE KPIs

	Voice over LTE key performance indicators
SIP drop rate	Session initiation protocol drop rate reflects that VoLTE drop rate due to fail to establish an IMS session
Voice quality/eMOS	VoLTE is typically measured as equivalent mean opinion score (eMOS) based on voice distortion level as reference to original voice sample or *perceptual evaluation of speech quality (PESQ)* to automate assessment of the speech quality as experienced by a user
Packet loss rate	Packet loss rates of voice digital samples, including DL/UL packet loss rate, max/average length of burst of packet loss, DL packet discard ratio
Latency and jitter	End-to-end voice transmission latency and jitter, RTP gaps
Throughput	DL and UL application layer throughput

We can also coordinate performance at eNB with the performance at EPC. For example, Fig. 4.9 shows with proper processing that application higher layer performance could be much better than that of physical layer eNB performance. Although physical layer's drop call is above 10%, the user application layer's drop rate is around 1.5%.

Operator can also obtain detailed network performance over all interfaces to ensure proper daily operation, such as potential overloading or malfunction of any interfaces. Soft defined network could utilize the information to redefine or partitioning network in real time.

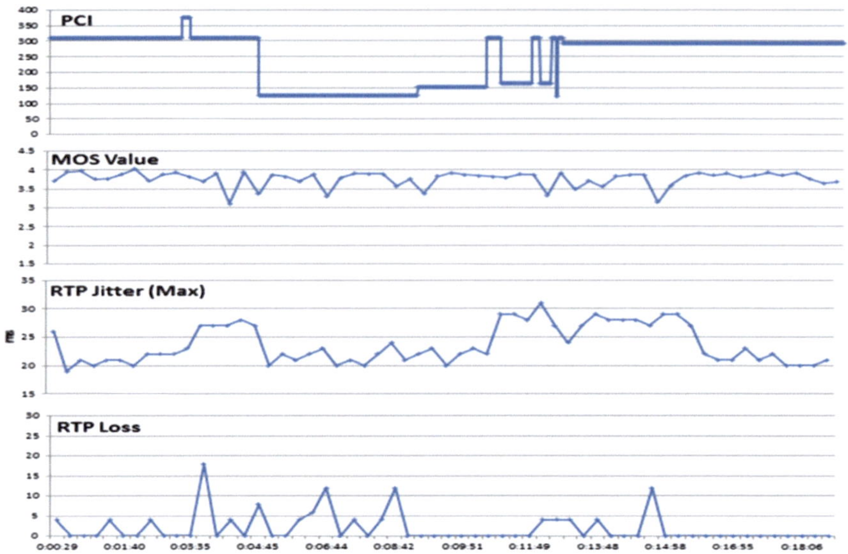

Fig. 4.8 VoLTE MOS versus packet loss rate and jitter

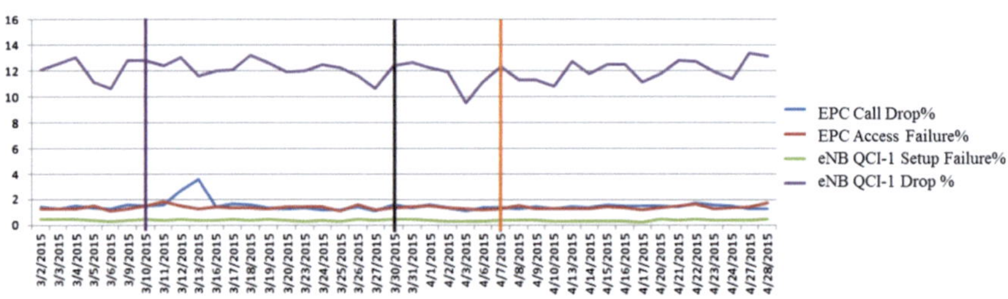

Fig. 4.9 VoLTE call drop measured at EPC and eNB

4.5 Summary

EPC network elements are dominant with software running on fast processors. This makes it first target to be virtualized.

3GPP has defined EPC architecture to support exponential traffic growth and mix-and-match of various network elements. Various interfaces and pooling capability among MME and SGW have flexible and dynamic routing capability to enable various applications and services. Soft defined networking and network function virtualization are enabling programmable and more adaptive EPC to various environments and services. Those established foundations to allow fine granularity and large scale of network performance and fault management models. EPC architecture presents much great opportunities for more advanced performance and fault analytic system to achieve optimized network performance and user experience.

Chapter 5
Overview of IP Transport Network Architectures—Backhaul, Metro, and Core Network

Acronym Key

CSR	Cell site router
EBH	Ethernet Backhaul
EDN	Engineering data network
eNB	eNodeB
IP	Internet protocol
LTE	Long-term evolution based on 3GPP standards
MLS	Multilayer switch
MPLS	Multi-protocol label switch
MSC	Mobile switching center
NID	Network interface device
PGW	PDN gateway z
SGW	Serving gateway
UE	User equipment

Many nationwide LTE wireless network service providers are challenged with providing coverage across massive geographical areas that can span hundreds or even thousands of square kilometers. Network architects first start addressing this issue by essentially breaking up a large network into smaller hierarchical segments:

1. **Core**—The main backbone interconnects various metros spread out across multiple geographical regions.
2. **Metropolitan Area (MAN) or Metro**—Typically, a ring network that spans tens to hundreds of square miles that interconnects backhaul networks to other backhaul networks via the metro, or via the core if across different metros.
3. **Ethernet Backhaul Networks(EBH)**—Typically a tree network that provides access for cell sites to other cell site, which may be local, or across different metros. This network segment is further broken into front haul that span femto cells, nano cells to provide more granular and wider coverage.
4. **Radio Access Network(RAN)**—This segment refers to the cellular network coverage area.

D. Kakadia et al., *Network Performance and Fault Analytics for LTE Wireless Service Providers*, DOI 10.1007/978-81-322-3721-1_5

The above list describes network segments in order of increasing complexity. The core is a relatively simple network, even though the core is carrying massive amounts of bulk aggregated traffic, the functional nature of the core is simple to move massive amounts of data across long distances. The metro has a little more complexity because we are connecting the EBH to the core or EBH to EBH, and there are several complexities where the art of designing the interconnections have trade-offs. The EBH is increasingly complex due to the tree architecture and the different types of connectivity involved which may be leased transport from local incumbent providers who may be competitors. The RAN is by far the most complex and hence detailed in a dedicated separate chapter.

Each of the above-mentioned network segments are optimized specifically for traffic patterns that it supports which may include any combination of characteristics: availability requirements, different protocols, levels of traffic aggregation, capacity, network element functions, and network measurement data.

The first step to developing network analytics is to collect the correct data for the entity to be studied. There are 3 fundamental elements of interest:

- **Network Elements**—These include network equipment or devices that move packets, bits, and optical wavelengths. The focus on measurement data pertains to the performance of the device itself in performing a specific function on a very small section of the end-to-end path.
- **Traffic**—This measurement focusses on the actual customer experience and at the end of the day is the most important measurement. The challenge is that this is usually the most difficult to actually measure in practice. Often operators take shortcuts and infer the customer traffic measurements by using synthetic active probes, which capture on a specific point in time, with the hope that the synthetic traffic injected into the network will experience the same performance as the actual customer traffic. There are limitations such as packet loss rates for example.
- **Transport**—This measurement concentrates on how well the service provider transport is providing a service that it has promised. For example, platinum or EF traffic has certain quantified availability, latency, and capacity. Measurement transport can be fairly accurately measured indirectly by inference by synthetic active probes since with the correct statistical distribution, reflecting actual customer traffic, with random samples and sufficient samples, decent p value accuracy can be achieved which is usually sufficient for transport network analytics.

The network analytics data can be further subdivided, for example traffic can be subdivided by protocols such as TCP Analytics or by function such as QoS analytics. Additionally, the data can then be cross correlated to provide key insights that impact the business in terms of improved customer experience, network optimization, and operational efficiencies by automating the sifting and processing of massive data in an automated manner to proactively predict potential customer impacting issues that can be detected and corrected prior to customers even being aware or impacted.

This chapter describes the main network segments of an LTE wireless carrier, followed by a description of how network analytics can be developed in the context of the network architecture subtypes. Most large-scale carriers have completely different designs and different network equipment and protocols for the different segment types which influence the network analytics that is implemented in each segment. As noted earlier, LTE wireless Carriers Transport Network architectures are composed of core, metro, Backhaul, and RAN. Due to complexity of the RAN, that is discussed in detail in a separate chapter of its own. This chapter will describe first the Backhaul network, then the metro and core together pointing out differences where applicable. The sections will provide the reader with an understanding of what measurements are important for the purposes of network analytics and finally at the end of the chapter discuss other important analytics topics that span these network segments including TCP analytics and QoS Analytics.

The following sections will describe in more detail each of the three network segment types and related network analytics that span the different geographical network segments.

5.1 Ethernet Backhaul Network Architectures

Backhaul networks are generally point-to-point connections from a mobile switching office to the cell tower. Historically, they were implemented based on T1 circuits. However, due to massive bandwidth requirements, backhaul networks are now implemented in DWDM optical rings. Ethernet backhaul network evolved from circuit-switched-based networks prior to the 2005–2007 time frame. Typical Backhaul networks were composed of multiple T1 links, each at 1.5 Mbps; however, as the RANs increased in capacity, the capacity grew dramatically on the backhaul and it was no longer cost effective to transport multiple Mbps using T1 circuits. Ethernet Backhaul offerings initially were actually MPLS VPLS services but as they further outgrew capacity, it became more cost-effective to run Ethernet directly over SONET, then recently directly over OTN. Most high-capacity Ethernet backhaul networks are based on the providers' dark fiber rings, where massive bandwidth requirements exist and investments warrant the high cost of dark fiber 20 year IRUs and availability.

Figure 5.1 shows how backhaul evolved from circuit-switched networks to Ethernet-based networks. At a time when carriers were using early GSM and CDMA where average download speeds were few hundred kilobits per second, backhaul networks of 1–3 Mbps capacity sufficed. However, as RAN technology advanced to HSPA and LTE, RAN downloads ranged from a few Mbps, the backhaul capacity requirements grew to 10–100 Mbps, and clearly circuit-switched backhaul networks no longer were able to scale at an economically feasible cost. Ethernet backhaul were needed scale beyond 100 s of Mbps.

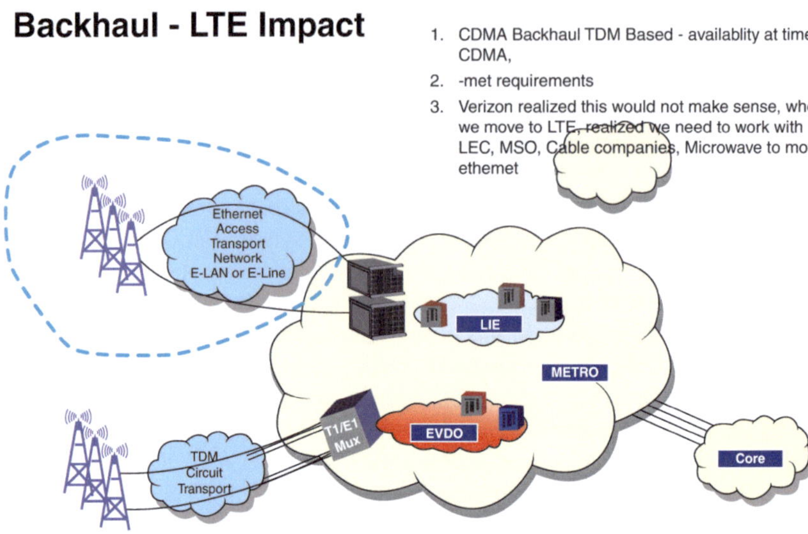

Fig. 5.1 Backhaul network types: circuit switched based, Ethernet based

Figure 5.2 shows how the early Ethernet service offerings were delivered by the LEC to the LTE wireless service provider. The LEC would have a regional MPLS network and easily provision VPLS or L2 VPNS to many customers in a LATA at reasonable cost. It is here where network analytics are needed for the LTE wireless Carrier to ensure they are getting promised capacity and to verify SLAs.

Figure 5.3 above describes the four main types of backhaul networks deployed by wireless carriers and each with different network measurement data and corresponding analytics. The following describes an overview of each type.

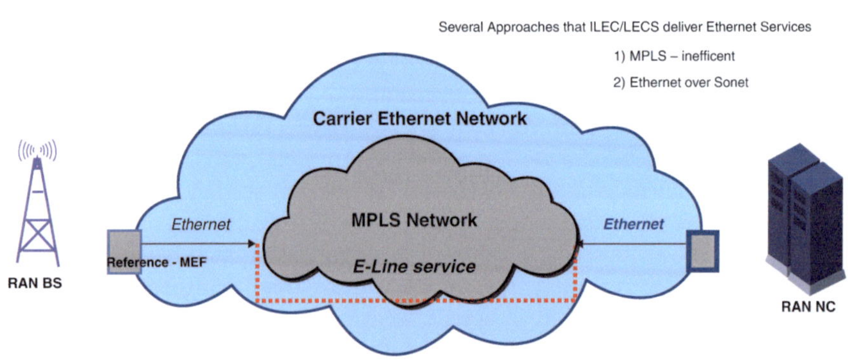

Fig. 5.2 Ethernet backhaul service offerings from LEC providers

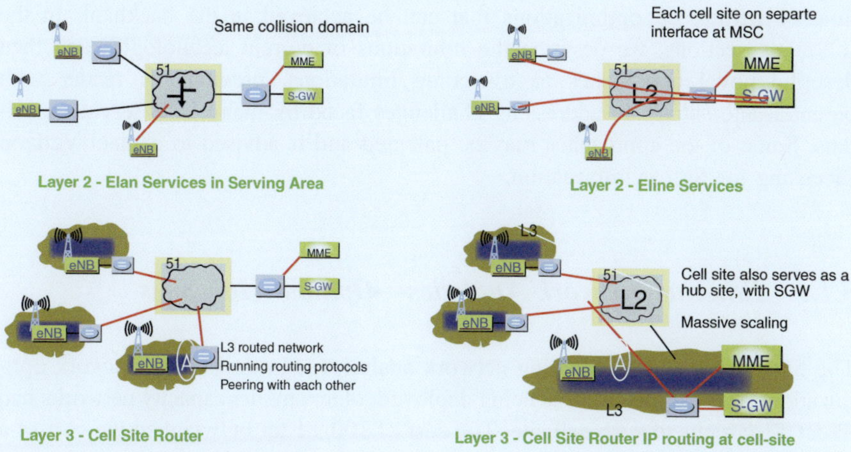

Fig. 5.3 Backhaul network designs

1. **Elan Services:** The wireless carrier is a customer of a LEC that provides an Ethernet Connection but the LEC uses an MPLS network with a L2 VPLS to deliver that service. This layer 2 connection will connect to other cell sites on the same broadcast domain. Elan services experience congestion since the LEC is sharing the underlying transport with many customers but also all sites will see broadcast packets on that same VLAN.
2. **Eline Services:** The wireless carrier is a customer of a LEC that provides an Ethernet connection but the LEC uses an MPLS network with a L2 Pseudo-wire provisioned to deliver that service. This layer 2 connection will often experience congestion since the LEC is sharing the underlying transport with many customers. This is different from ELAN services in that these are point-to-point connections and broadcasts can be controlled by the wireless carrier at the router or switch at the MSC.
3. **Layer 3 Cell Site Router:** The wireless carrier is a customer of a LEC that provides an Ethernet connection but unlike ELAN and ELAN which only require a low-cost L2 Switch, in this case a Layer 3 is required to support routing to multiple networks through static route configurations. This is adequate for smaller scale deployments.
4. **Layer 3 Cell Router with IP Routing:** This type of backhaul of network is similar to above but instead of static route configurations, the cell site router and MSC Ethernet Backhaul router exchange IGP or EGP routes. This is suitable for large-scale deployments. The network analytics will be the most complex in this case.

Now that the reader has a fundamental understanding of the Ethernet Backhaul network segment, we will now describe some innovative network analytics that can be applied to a backhaul that provides the operator with unprecedented insights, at time of writing on optimizations that can be achieved in the backhaul. In the following sections, we describe the limitations of current technologies and then describe novel approaches to overcome limitations, providing a reader with potential alternatives to addressing challenges faced by many LTE service providers. Some of the approaches may be patented and is advised to contact Verizon Licensing for further information.

5.1.1 *Ethernet Network Analytics—Optimal Hub Sites*

The first application of applying network analytics is in the area of network optimization. At the time LTE was being deployed, older smaller capacity networks had direct T1 circuits to each cell site. The cost of 100 T1 far outweighed the cost of a 10 Gbps Ethernet circuit, compounded the benefit of stat mux, huge savings, and better capacity could be achieved. The problem was to identify the optimal hub sites, as there are many possible choices.

Figure 5.4 shows an example of network analytics used to optimize the Backhaul network, where a hub aggregation architecture was analyzed in terms of latency and costs. The green lines show less than 25% link utilization, blue is 50%, and red (not seen) is 90% congestion. Costs are also included in the network analysis and the optimal architecture in terms of performance and cost based on geographical locations of the cell sites, MSC can be computed. This has the potential to save the carrier millions of dollars per year with better performance than

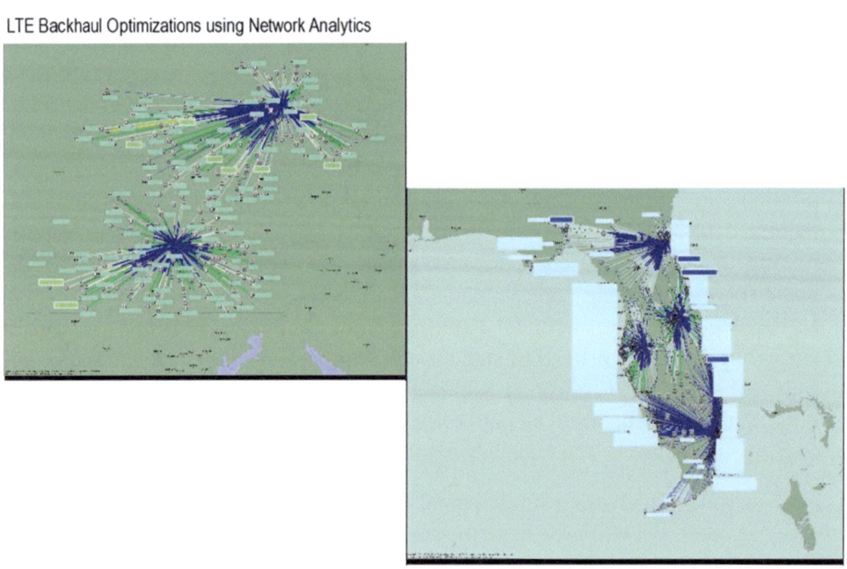

Fig. 5.4 Example Ethernet backhaul network analytics and optimizations

without analytics. The approach involved finding the optimal hub sites, where the root of the tree should aggregate traffic from the cells to the MSC.

5.1.2 Ethernet Backhaul Network Analytics—Measuring Actual Customer Experience

At time of writing, current technology to measure the quality of a service provider Ethernet Backhaul network is highly inaccurate and does not measure the actual performance experienced of actual customer traffic. Instead a guess-based approach is used where synthetic probes are injected into the network and based on that synthetic traffic, it is assumed that the customers traffic are similar in terms of quality. The fact is that actual customers traffic can take completely different routers by a backhaul provider, and backhaul providers often are able to determine synthetic traffic from actual customer traffic and provide artificially good treatment on the synthetic traffic.

Current methods to assess the quality of an Ethernet backhaul (EBH) link do not provide data that is granular enough for an overall assessment of customer experience or troubleshooting end-to-end performance. This section describes a novel approach to gain highly granular EBH performance monitoring data by segmenting the network analytics. Specifically, data collectors are strategically placed throughout the network to collect EBH performance KPIs per UE, per bearer, or per network segment and then aggregated in a centralized analytics engine for correlation.

Figure 5.5 illustrates the current method for assessing the quality of an EBH link. The quality of the EBH link is tested by a synthetic packet usually every five minutes. Limitations of this traditional approach are explained as follows. The synthetic packet is different from the actual customer traffic which is what we are interested in measuring. The synthetic packet, for example an ICMP ping, will experience different network latencies than actual customer traffic, for example a video stream of MTU sizes. The user-perceived quality of the link measured using the synthetic packet is not accurate. The actual traffic is usually experiencing a different type of latency; since as links are getting faster, millions or billions of packets are not measured and many episodes of congestion are not measured. The proposed approach measures actual traffic every 0.1 ms and computes the departure rate minus the arrival rate to gauge the latency across the backhaul in a highly precise and useful manner. Common KPIs are also calculated across the link and used to enrich the latency data as well as report to a user interface for analysis by network engineers.

It is very difficult to measure actual customer traffic in a non-intrusive way, in particular the actual network latencies, jitter, packet loss, and other metrics actually experienced by customers, whose traffic actually rides on a certain backhaul segment. Most instruments only see bits and bytes and often cannot accurately measure

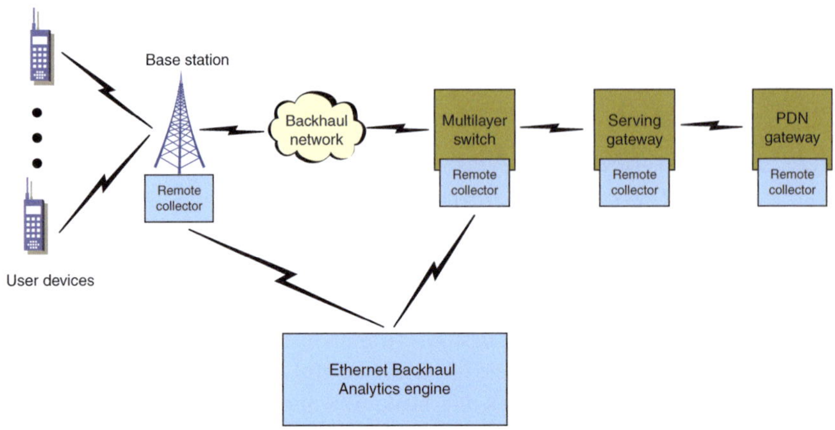

Fig. 5.5 Ethernet backhaul network analytics—measurements based on synthetic data versus actual data for increased accuracy

at high speeds. Using an analytics approach, we describe a novel approach that can actually measure the actual customer experience on the backhaul, where the data is processed, analyzed, and only anomalies are escalated to reduce the amount of noise that an operator must process and focus on the important issues. The analytics approach uses a multi-tiered approach where real-time, mid-time, and long-term trends are processed and the resulting data is processed in columnar databases for rapid report generation.

During off-peak hours, the UE will generate test traffic. As this traffic passes through the network, its time stamp will be recorded by a remote collector in a table format. The data will be gathered through a passive span, as represented by the dashed lines in the figure. Other information will be recorded by the Remote Collector including, but is not limited to:

- Source IP address
- Destination IP address
- Source port
- Destination port
- Packet ID
- UE bearer ID
- EBH performance KPI.

The remote collectors will be strategically placed throughout the network path in order to measure specific segments. Segment 1 will be identified as the RF link between the UE and the eNB. Segment 2 will be identified as the EBH link between the eNB and the MLS. Segment 3 will be identified as the network link between the MLS and the SGW. Segment 4 will be identified as the network link between the SGW and the PGW. Each remote collector table will be compressed and then aggregated in a centralized analytics engine. IPFIX information from the PGW will also be sent to the centralized analytics Eengine. This information allows for the data bearers to be correlated to an IP address. The centralized analytics engine will calculate the network performance KPIs per UE, per bearer and per segment and will then process this data for a user interface.

The remote collector will use to collect and compress the data for transmission. When the process starts, the remote collector will assess if the data is being generated from the PGW. If the data is from the PGW, the collector will begin building a remote collector table for the identified segment with the time stamp, IP address, bearer and KPI information. If the data is not from the PGW, the collector will begin gathering packets for analysis. The time stamp for the individual packets is recorded. The bearer ID, protocol data, and packet ID are extracted from the individual packets. This information is then used to populate a remote collector table for the identified segment. Multiple bearer IDs and KPIs can be recorded for a single time stamp. The time stamps and KPIs are then compared against user-defined thresholds. Any threshold violations will be processed and then sent to the centralized analytics engine for alarming purposes. The entire remote collector table will then be compressed and forwarded to the centralized analytics engine.

The centralized analytics engine will use correlate the data and prepare the data for analysis. The centralized analytics engine will begin the process by collecting the remote collector tables from each device. Each table will contain the information collected from each network segment. The engine will then build the necessary analytics cubes. These analytics cubes will compare the data in multiple dimensions and will provide a more accurate assessment of the end-to-end network performance quality. Finally, the data will be formatted for a user interface and presentation graphs.

A collective testing method can be implemented that is scalable. The current solution requires a BV box to be deployed at each MSC in order to test the EBH links of multiple cell sites, as represented by the blue boxes in the figure. The proposed solution will allow for the Y.1731 testing to take place in a more centralized location. A single testing platform will be deployed at an area level and will still be able to identify more granular network segments, as represented by the red box in the figure.

5.1.3 Backhaul Analytics Insights

Consider Fig. 5.5 again, which also shows an example how the Ethernet Backhaul measurements that are taken and processed to determine not only the quality as experienced by subscribers at the Ethernet segment, but also insights into the quality of the transport service provided by the LEC Ethernet backhaul service provider. Based on the data collected, valuable analytics and insights can be deduced. There are 3 modes of analytics:

1. Real Time: Remote computations as the data is streamed in can be computed as the data is collected, as soon as certain thresholds are exceeded, an immediate real-time analytics computation can be performed to alert the service provider operator of any potential issue. This can be implemented using existing Hadoop technology using the map request computations facility.
2. Midterm: As per cell site data is aggregated into regional computations are gathered, further correlations and analytics can be computed based on a larger set of data sets, for example all cells sites on a per regional basis. This combined with a midterm time span, say for example 15 min to 1 h, a wider range of analytics and insights can be gathered to provide the operator further deeper knowledge on how well the Ethernet backhaul is performing as compared to other cells within that region or across backhaul providers if multiple in the region. Analytics can be segmented by region, backhaul basis and the analytics can be computed for that particular comparisons.
3. Long Term: as regional data is further aggregated into a nationwide computation, further correlations and analytics can be computed to provide the operator further insights based on a per region basis or backhaul provider basis to provide more holistic comprehensive insight on a nationwide basis (Fig. 5.6).

Overall, the proposed method for monitoring the end-to-end network performance quality through the segmented analytics offers several benefits:

1. Accuracy: The proposed solution offers a higher level of accuracy than current methods by measuring the network performance of actual customer traffic. Current methods use test data to measure the network performance, which only simulates and estimates how customer traffic is treated.
2. Packet Loss Identification: As the proposed solution tracks the Packet ID through each segment of the network, where a packet is lost can be identified. This information would be valuable when troubleshooting a customer issue and could assist in identifying chronic troublesome links.
3. Latency Identification: As the proposed solution tracks the time stamp of each packet through each segment of the network, the latency each packet experiences can be calculated. This information would be valuable when troubleshooting a customer issue and could assist identifying chronic troublesome links.

Timestamp	IP Src	IP Dest	Port Src	Port Dest	PID	Bearer ID	KPIs	
t1	1.1.1.1	1.1.1.2	21	8892	Ox0101	Ox01	●●●	●●●
							●●●	●●●
t1	2.2.2.2	2.2.2.3	20	1234	Ox0202	Ox02	●●●	●●●
							●●●	●●●
t2	3.3.3.3	3.3.3.4	88	5555	Ox0303	Ox03	●●●	●●●
							●●●	●●●

Fig. 5.6 Ethernet backhaul data measurements—analytics provide insights on actual customer experience and EBH provider SLA compliance

4. Scalability: The proposed solution would allow a test unit to be deployed at an Area level rather than at each MSC.
5. Nationwide Analysis: Only KPIs that cross a pre-defined threshold will be transmitted for alarming. This reduction in traffic allows for only high-valued targets to be viewed. As a result, a nationwide analysis can be executed to determine how the entire network is performing as a whole.
6. Jitter Source Identification: As the proposed solution tracks the common EBH KPIs through each segment of the network, where a packet is experiencing jitter can be identified. This information would be valuable when troubleshooting a customer issue and could assist in identifying network bottlenecks.
7. Throughput per Segment: As the proposed solution tracks the common EBH KPIs through each segment of the network, the throughput per segment can be calculated. This information would be valuable when troubleshooting a customer issue and could assist in identifying network bottlenecks.
8. Fragmentation Identification: As the proposed solution tracks the common EBH KPIs through each segment of the network, where fragmentation of a packet occurs can be identified. This information would be valuable when troubleshooting a customer issue and could assist in identifying chronic troublesome links.

5.2 Core and Metro Network Architectures

In this section, we examine core and metro network architectures together due to the constant changes in technology, the components used in the core are almost always used in the metro over time. The term core network or backbone is characterized loosely be spanning the entire service providers network, interconnecting the next lower aggregated network, which is usually the metro network for a nationwide carrier. The term metropolitan area network (MAN) can be defined as a collection of connecting local area networks (LAN) and geographically span a few kilometers to tens of kilometers. However, in the wireless service provider industry, the term MAN or “Metro Network” is loosely defined as the high-speed optical transport network that connects mobile switching center (MSC) that span several cities usually 4–20 max in a ring topology. MAN technology has evolved from DS3 (45 Mbps) which carried mostly circuit switched voice traffic to SONET (OC3 to OC 192) which carried mix voice and data to Ethernet over OTN (ODU 2 10GE to ODU 3 100GE) which now carries predominately data. Optical technology made significant technological strides in the 2000s decade, one of the most important being the introduction of the Reconfigurable Optical Add Drop MUX (ROADM). Technological advances in long range optics and lasters that eventually led to ultra-long distance which allowed the service provider to transport massive amounts of data at relatively low costs over relatively long distances. This section will describe the MAN basic network architecture and then how network analytics can be used to optimize the transport of traffic and perform reconfigurations based on analytics. The proposed MAN network analytics can be operationalized as a standalone system for use by network operations, network engineering, or network planning, and more recently can be incorporated into and SDN controller. MAN network analytics can provide the SDN controller with very accurate information on how and when to reconfigure the MAN network based on monitoring and analyzing how parameters can be tuned to meet certain performance objectives which optimizes goals such as maximum QoE, maximizing network utilization, maximizing performance, or a combination of objectives.

The metro network aggregates hundreds to thousands of Ethernet Backhaul measurements from a network analytics perspective. From a functional standpoint, the Metro network interconnects MSCs, and each MSC connects to hundreds to thousands of cell sites. Figure 5.7 shows a high-level description of a four-tier model of a typical LTE wireless service provider network architecture which consists of the following segments or tiers:

- Core
- Access/Distribution
- Ethernet Backhaul
- RAN.

The access/distribution layer of this diagram is deployed on the CDMA or GSM wireless service providers metro network. Typically, Access locations aggregate traffic from tens or thousands of remote cell sites, and these Access

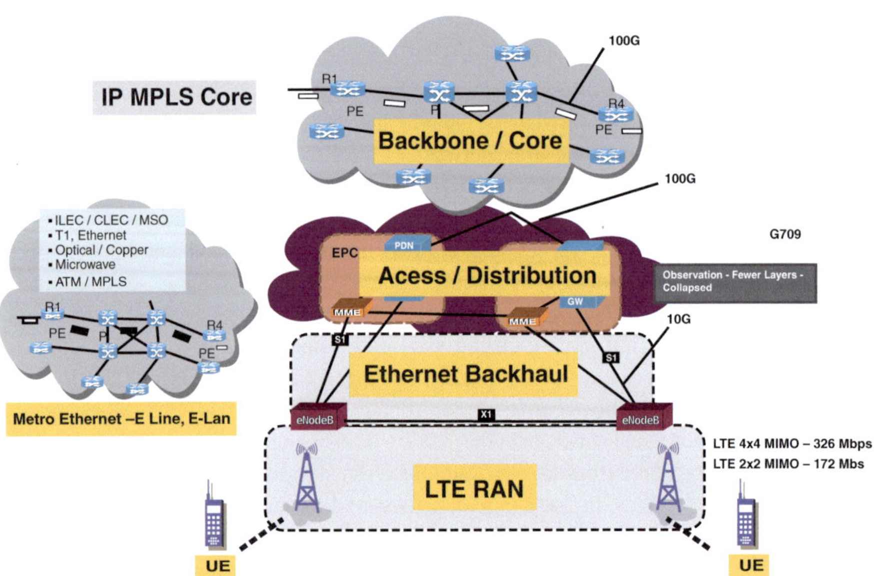

Fig. 5.7 Network model of example LTE wireless service provider

locations are interconnected by the wireless service providers metro Network. The metro Networks typically DS3 had to transport very little traffic since the bandwidth of CDMA and GSM was much less that LTE. Due to limited optical switching technology, metro networks were connected by regional metros which then connected to a large bandwidth core. One of the elements in the metro is the distribution location which aggregates all the traffic to the core, or if Internet Data, can hand off the Internet traffic to peering point co located at the distribution location. Often there is confusion on a second metro Network which provides circuits for the LTE service providers Ethernet Backhaul links. In different parts of the country, there are different local exchange carriers (LECS) or other providers that have resell connectivity to enterprises in that region which is transported on the LECSs metro Network. This book will not discuss these metro Networks as they are not a part of the LTE service providers network. However, in reality, the LTE service providers often are related to a larger company that also provides wireline transport, so the LTE service provider may also be leasing Ethernet Backhaul circuits from its sister company which both belong to the same parent company.

Figure 5.8 describes an example LTE wireless carrier network architecture showing how the Metro Core and Metro Regional connect access and core. The Regional Metro is used to scale several smaller metros.

In the early 2000s since the main traffic was voice, and Internet traffic was not prevalent in the CDMA or GSM wireless industry, there used to be thick large core backbone and small metro networks. The voice traffic patterns were aggregated to the core and Internet traffic was often backhauled to very few peering points in the

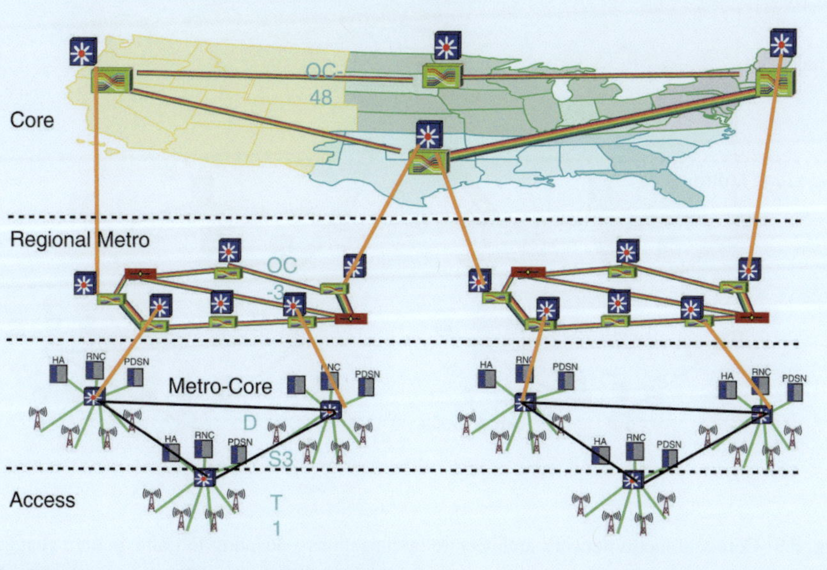

Fig. 5.8 Example 4-Tier pre-LTE wireless service provider network architecture—metro interconnects several access/backhaul networks

LTE service providers network. However, from the latter half of the 2000s decade up to 2014, Internet traffic has dominated the traffic patterns and LTE service providers typically have distributed peering points as close to the user as possible in order to reduce costs required to backhaul Internet traffic. This resulted in a dramatic shift of the LTE service provider network where the metro Networks are massive and the core is small. This can be seen in Fig. 5.9 where not only the metro is thick, but also the four-layer architecture collapsed to three layers, by removing the regional metros.

Metro Networks connect several geographically nearby local access networks to a core. Figure 5.8 above shows an example of a typical network architecture showing how these networks are related to each other. This architecture was originally designed for a mixture of circuit-switched-based voice and low bandwidth data of that period in time. As LTE offers relatively much more bandwidth at the RAN segment, this causes a ripple effect on the metro adding up traffic from tens to thousands of cell site traffic. This architecture is clearly not optimal for LTE service providers due to the higher bandwidth requirements which predominantly must be handed off to the Internet, starting from the access to distro sites. In 2000–2005, the access, Metro core were typical OC-12 to OC-48 SONET connections which did not scale, not only in terms of bandwidth, but surprisingly on scaling the cost of transit traffic. Metro cores are typically rings, where access locations forward traffic to a distribution location, which is essentially 1 node on that metro ring.

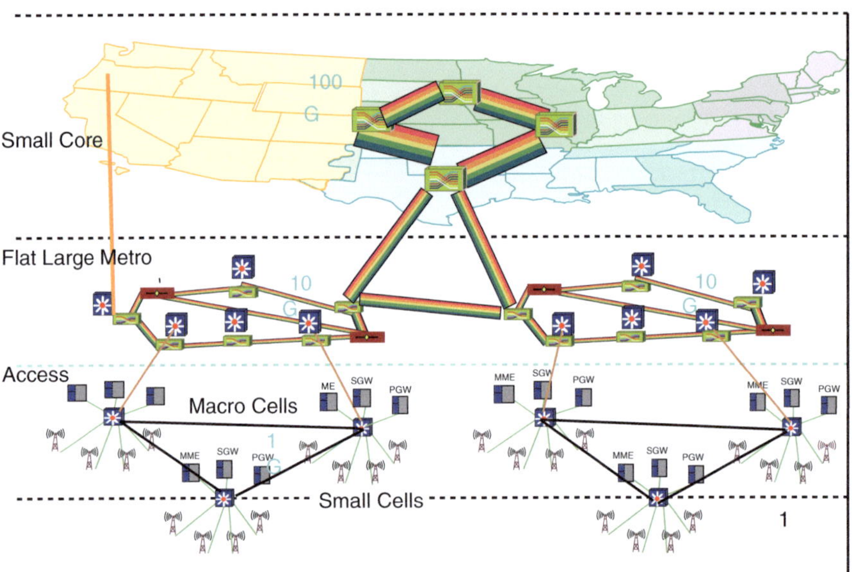

Fig. 5.9 Core and metro network architecture optimizations—adapting to traffic pattern changes

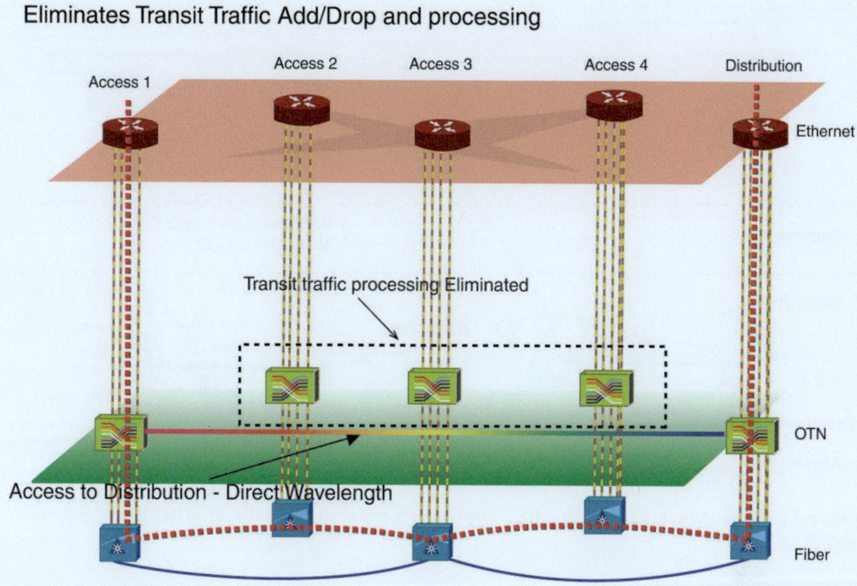

Fig. 5.10 Metro and core scaling by using optical bypass at intermediate hops

Intermediate access locations had to process transit traffic at layer 3, which requires router processing and forwarding. Figure 5.10 shows how as traffic increases, the amount of traffic processed at each intermediate hop (not shown) would require traffic to be processed at each intermediate Layer 3 router and transit accessed location increases additively since as more traffic is originated further away from the hub or the distribution site, more intermediate access transit nodes must not on process but also forward to the upstream neighbor, thereby forcing more traffic on both interfaces and line cards. This design was highly desired by router vendors who would reap massive profits on the interfaces, which operators only realized after it was too late. The authors during their tenure, came up with a far better design, that served the interests of the operator, which was to bypass traffic at each intermediate hop and keep the traffic in the optical domain, eliminating the need for not only unnecessarily dropping the traffic, but also reducing load and improving performance by saving transit delays.

The solution came from the introduction of ROADMS, which replaced ADMs. ADMs were designed for SONET and as G.709, Ethernet directly over OTN emerged, ROADMS also added massive scale benefit to LTE service providers because transit traffic would be forwarded in optical domain, at no extra transponder cost, and no additional transit load on the access transit routers. Figure 5.10 shows

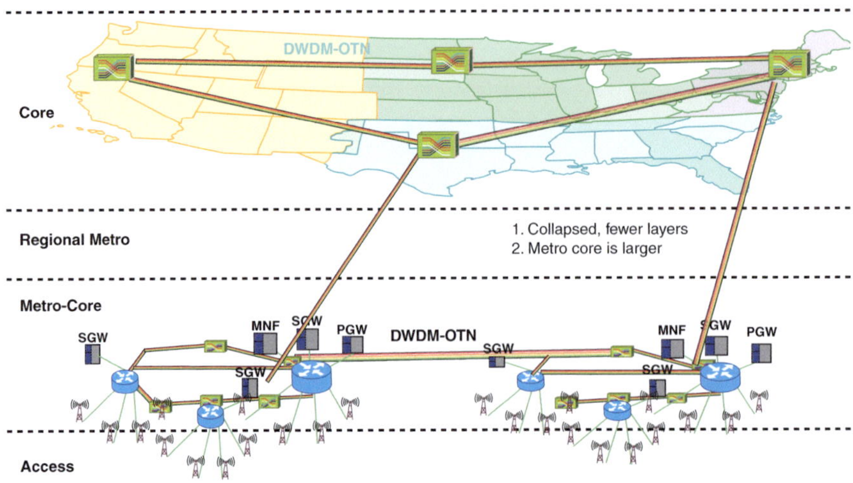

Fig. 5.11 Next-generation metro—flat large metro resulted by collapsing regional and metro networks

the improved metro architecture using ROADMS. Network analytics was used to justify the improved performance and lower cost. In the original design, the traffic processed at each intermediate hop from originating access to distribution site grew; as you traverse intermediate hops, the hops closest to the distribution site carried the most traffic. Interfaces on the routers were a huge cost. With the hub and spoke, network analytics clearly showed for the same traffic patterns costs were reduced significantly and performance improve due to the elimination of intermediate traffic processing, and associated additional latencies.

As a result of the deployment of ROADMs in the metro, and the fact that traffic patterns changed from predominantly voice to Internet data, where each distribution site had Internet handoff, the result next generation architecture resulted in a smaller core, and collapsed metro and regional metro to a single, high-capacity metro optical network as shown in Fig. 5.11.

In order to understand insightful network analytics, one needs to understand the metro from high to low levels. Figure 5.13 shows a high-level abstract typical metro Network architecture diagram showing only the logical connectivity between access locations within metro. Figures 5.12, 5.13, and 5.14 show further increasing levels of reality and granularity. In Fig. 5.12, a packet going from San Francisco to Fremont looks straightforward; however, we see in Fig. 5.13, the circuits involved are more complicated and Fig. 5.14, which shows the layer 1 fiber connections in increasing complexity and often inefficient paths to take compensate for nonexistent fiber runs.

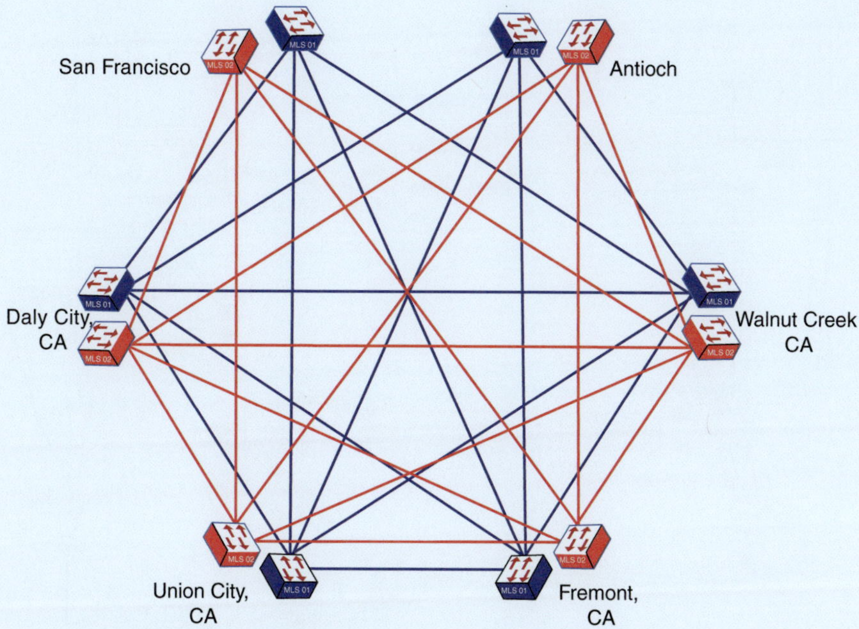

Fig. 5.12 High-level logical IP metro full mesh connectivity diagram

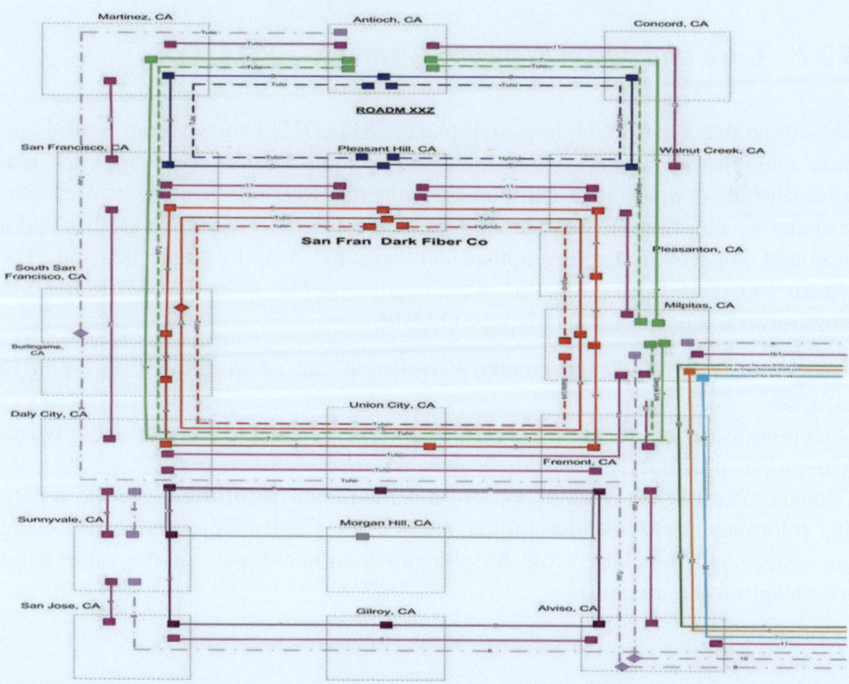

Fig. 5.13 Increased level of detail of metro connectivity diagram—circuit diagram

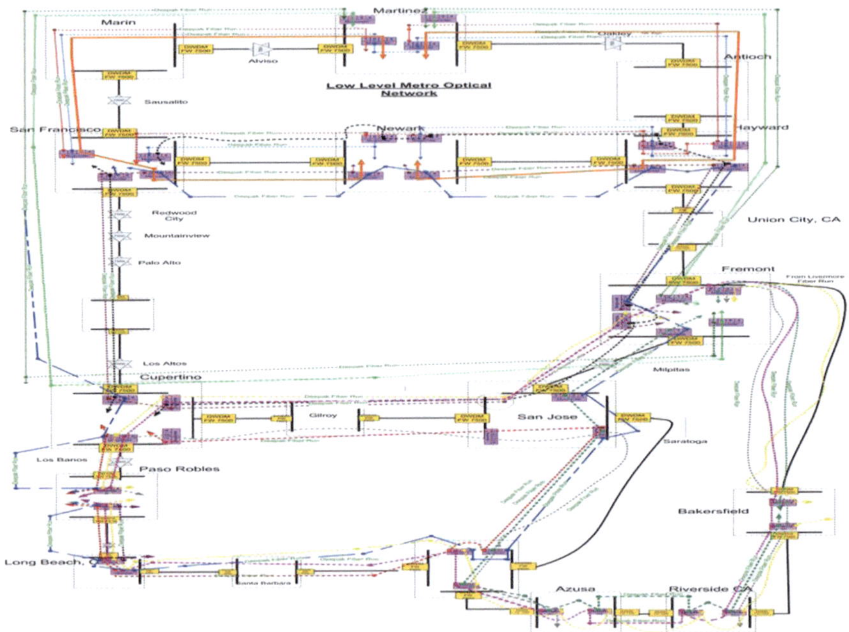

Fig. 5.14 Low-level metro network connectivity showing inefficiencies—fiber diagram

5.2.1 Core and Metro Network Elements—ROADMS

Reconfigurable optical add/drop multiplexer (ROADM) evolved from optical add drop multiplexers which is fundamentally an optical cross connect switch that instead of fixed wavelength and port assignments, ROADM can be software provisioned to add, drop of transparently pass thru a node. This flexibility has had a profound impact on adapting optical networks to changing traffic demand. The typical ROADM is based on the wavelength selective switch(WSS). ROADMS allow several advantages over legacy OADM.

Colorless: Using WSS, any colored wavelength can be assigned to any port in software.
Directionless: Using WSS, any wavelength can be routed to any port in software, without physical intervention.
Contentionless: When a wavelength of a certain color is to be assigned to a port, that color may already be in use for another circuit. Contentionless refers to solving this wavelength blocking issue by allowing multiple copies of the same color wavelength to be assigned

5.2.2 Core and Metro Network Elements—IP/MPLS Routers

In this section, we describe the IP and MPLS building blocks, dominated by Cisco, Juniper, and now with new vendors Alcatel-Lucent or Nokia and Arista.

The MPLS router comes in two forms:

- **PE—Provider Edge**, this is the most important MPLS router, it is where the MPLS demarc begins.
- **P—P routers** allow MPLS networks to scale, since each PE router must be connecting to each other, by adding P routers, the network can scale by reducing direct PE to PE connections. P routers create a hubbing effect to reduce the number of links needed to provide full PE to PE connectivity.

Typically, IP MPLS provide layer 2.5 and 3 connectivity in the metro as well as providing VPN connections for external customers, which is not common for LTE wireless service providers to provide, but often are customer for other LEC providers in the Backhaul networks.

5.2.3 Core and Metro Transport Links

Transport links in the Core and Metro can be monitored at various layers to pinpoint source of congestion and faults. At the highest network layer 3, typically BFD pings are used to verify link integrity and latencies or inferred congestion. However, as links become faster, these approaches do not scale and can be areas of further research.

Optical performance monitoring (OPM) has recently been advancing due to the importance of the optical transport layer supporting increasing amounts of revenue generating customer traffic now reaching multiples of 100 Gbps links. Recent advances in OPM allow the operator to identify and predict optical layer faults, diagnosis, and localization and take corrective action in the event a fault is predicted or increased performance can be achieved by tuning parameters for compensating for optical impairments due to for example changes in temperatures or failing amplifiers. OPM includes monitoring analog or digital parameters of the transmission system. There are basically three elements that can be monitored.

The optical signal: This includes the following parameters that can be monitored:

1. Optical Signal-to-Noise Ratio (OSNR)—using some form of spectrum analyzer, array wave grating, or tunable filters, the ratio of noise to the original signal is measured.
2. Chromatic Dispersion Monitoring (CD)—one technique involves measuring at the receiver, the power variation of different sub-carriers due to fading resulting from dispersion.

3. The originating device of the optical signal, practically this refers to the amplifier.
4. Fiber Link: This includes the following parameters that can be monitored.
5. Optical Time Domain Reflectometry (OTDR), monitoring point-to-point fiber runs with a pulse sent and received and measured.

5.2.4 Core and Metro Network Analytics and SDN

In the previous sections of this chapter, we reviewed the Core and Metropolitan Network layer by layer. In order to build a network analytics framework that can be used to develop analytics applications, we need to identify the goal of the analytics results. The existing network performance management tools commonly available from commercial vendors simply take counters and chart them, where more advanced tools do some linear extrapolation of the data to provide some degree of forecasts. We propose some ideas that are used to provide insights that are not obvious and can be used for future work. The following sections will describe the network analytics strategies and examples of how to develop the performance models for that approach.

5.2.5 Multilayer Network Performance Correlation

Core and Metro networking protocols are based on layers. A lower layer network module provides a service to a higher layer. For example, TCP is a transport layer protocol that provides a reliable, in order transport service to a socket application that wants to send a byte stream across the network to another endpoint application.

Network performance data and analytics can be used on an ongoing basis for a network operator to continuously determine what the network availability is at any point in time. This is a very important measurement since the network is continuously changing and in order to determine the feasibility of spending money by adding links or more expensive routers and figuring out what the overall benefit to the company is in terms of network availability is a very difficult task. This section will propose an approach how continuously instrument network performance measurements and configurations can be used to determine overall network availability. This will require crossing several different domains of knowledge such as network design, configuration management, and architecture.

At time of writing, it is difficult for a network operator to accurately determine what the actual availability is on the network that also considers that changes in the network. Using the multilayer network analytics proposed, this can be done in a programmatic fashion. The key is in computing the overall network availability considering the latest state of each layer of the multilayer/multi-domain networks, thus computing the overall network availability based on an incremental change such as addition of an expensive link that does not add any real benefit to the company since the incremental value is marginal compared to the cost. As today's end-to-end network is cascaded with networks with different layers such as IP, MPLS, Ethernet, SONET, and WDM, the method of computing the highest network availability to multilayer/multi-domain networks is urgently needed. Compared to the existing method of computing the network availability of each layer network independently, the proposed method will result in lower cost of operations with the same target network availability for the overall network as a system.

Wireless service providers typically have a minimum level of network availability requirements; for example, 99.999% of network availability, or less than of the probability of network (PF) failure.

The design of networks typically involves different teams with different skill sets, which include, IMS services, data, and transport groups. In practice, different teams work independently and efforts are not aligned. The result is that some layers of the network (e.g., SONET and WDM networks) are over-engineered and other layers (e.g., IP/MPLS networks) are under-engineered, resulting in an overall system wide network that is sub-optimal. The network is sub-optimal, because there is no alignment of efforts to ensure that the overall network as a system is designed with the target network availability. As today's end-to-end network is cascaded with networks with different layers such as IP, MPLS, Ethernet, SONET, and WDM, the method of computing accurately network availability to multilayer/multi-domain networks is urgently needed. Compared to the existing method of meeting the network availability of each layer network independently, the proposed method will result in lower cost with the target network availability for the overall network as a system.

The design of MAN network architectures for wireless service providers involves the design of networks at the following layers:

1. **Service Control Layer**—this layer is designed to support control signaling traffic, which has the characteristics of low traffic volume, but at the highest priority.
2. **Service Bearer Layer**—this layer is designed to support voice, video, and data traffic, which has the characteristics of low to high traffic volume and at various different priorities. Data traffic can tolerate delays, and even some packet loss, relying on retransmission at the sender side, or very low tolerance to latencies and jitter for voice traffic, but relatively low volumes of traffic at the subscriber level, but can be aggregated to tremendous volumes of voice data at aggregation points.
3. **IP Control Layer**—this layer is designed to support routing and other control traffic used to inform routers about the paths and routes in the network and other state information.
4. **IP Data Layer**—this layer simply forwards IP traffic.

5. **MPLS Control Layer**—this layer is designed to support control traffic for MPLS, which includes signaling RSVP to set up traffic engineering tunnels, distributing labels to adjacent MPLS routers using LDP or TDP.
6. **MPLS Data Layer**—this layer is designed to forward MPLS traffic based on the MPLS Label.
7. **Data Link Layer**—this layer provides the framing and access to the media, using Ethernet or other data link layer protocols.
8. **SONET**—this layer provides the optical channels to transport the data link layer frames.
9. **WDM**—this layer transports SONET on wavelengths.
10. **Fiber**—this is the physical optical layer.

Figure 5.15 below shows the network layers in a typical wireless service provider IP MAN transport network

It is important to note that many of the layers, such as IGP, BFD, MPLS, and SONET will have its own availability mechanisms, which run with no knowledge of each other, in effect duplicating availability efforts. Further, some of these efforts will not add any benefit since there are cases where the backup path will actually be transported on the same fiber that failed, and if the fiber cable breaks, then so will be both the primary and backup paths.

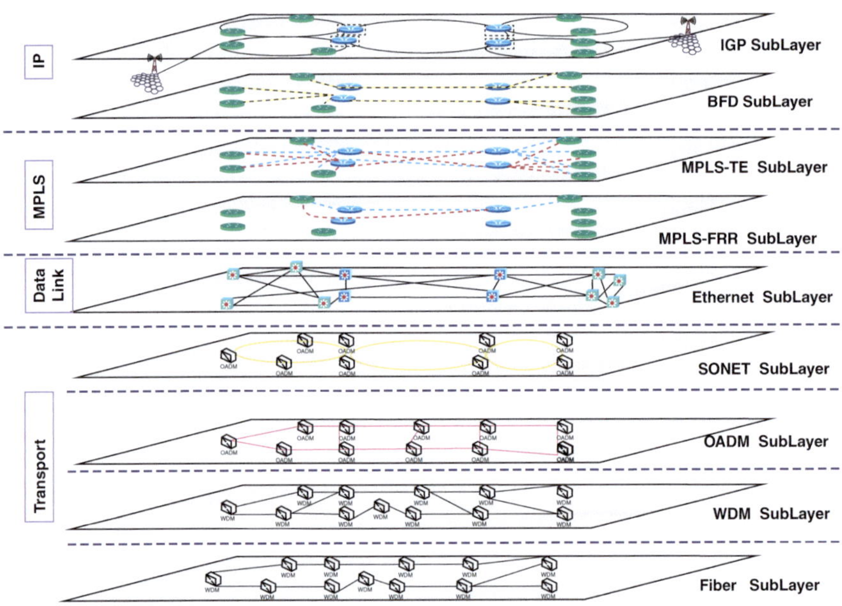

Fig. 5.15 Typical wireless service provider network architecture decomposed into various network layers for metropolitan area networks and core networks

What is needed is a methodology, which not only considers all layers of the network architecture, but also that each layer is designed with availability mechanisms that are aligned with the other layers, such that the net overall system result meets the target system availability requirements.

In this section, we will illustrate the traditional method to calculate the availability of a particular network architecture. For simplicity, we will show only a partial computation to make a point. In the following sections, we will show enough of the computation for the purpose of comparing with the proposed approach.

For illustrative purposes and without loss of generality, we take simplified network architecture as shown in Fig. 5.16. The diagram represents an abstraction that represents how a packet moves from source to destination and the layers of protocol processing involved in this journey. A packet will traverse each layer up and down the protocol stack many times at different physical locations until it reaches its final destination.

Another view on how a packet moves from source to destination via different layers is shown in Fig. 5.17. In general, packets must traverse many different layers in transit from source to destinations—layer 1 of SONET or WDM, layer 2 of Ethernet, ATM or Frame Relay, layer 3 of IP and so on.

Application engineers design into the application availability mechanisms such that if the primary service fails, which is identified by a UDP or TCP socket connection, a backup service that will be available, using either an alternate or backup UDP or TCP socket connection or some type of similar load balancing mechanisms. But IP network engineers will design availability mechanisms at the

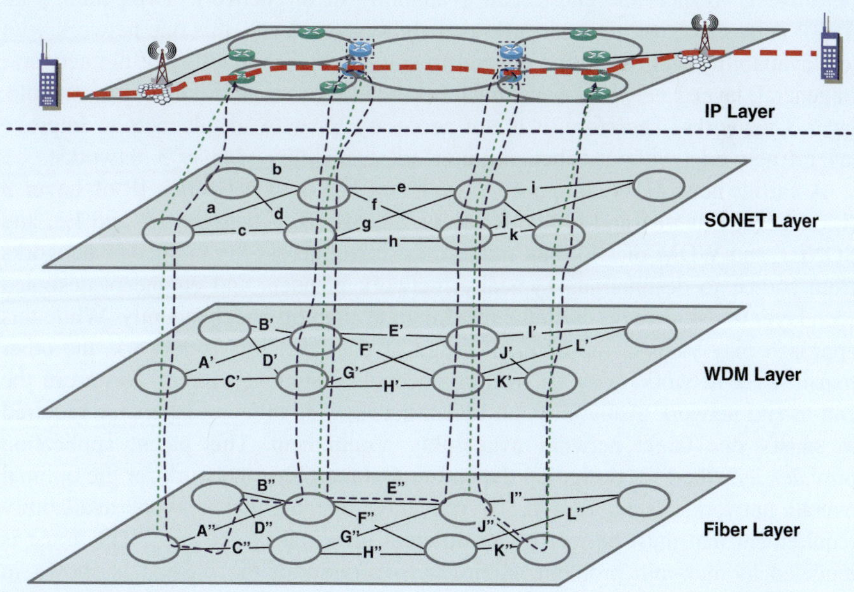

Fig. 5.16 Simplified network architecture—4 layers

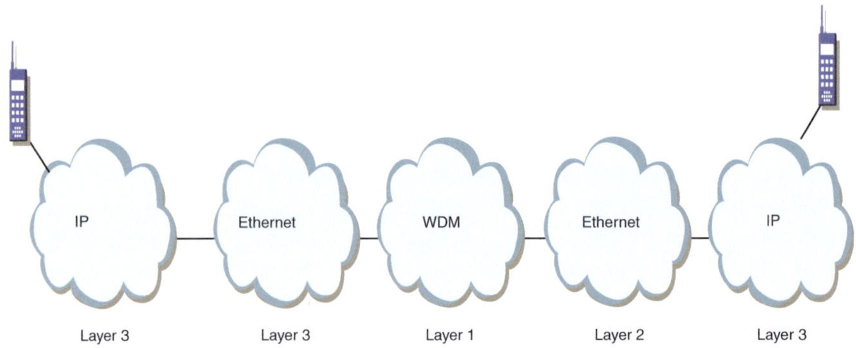

Fig. 5.17 Packet traverse of 4 layers

IP layer, without any knowledge of the availability mechanisms already provisioned at other layers. Similarly, optical transport engineers will also design availability mechanisms at the SONET layer, WDM layer, and possible fiber layers, independently of the availability mechanisms already designed at the higher network layers. The result is often over-engineering at some segments, and insufficient availability at some other segments.

The following method for network fault analytics is based on patent technology which the author was a contributing author.

The current method of meeting target availability is meeting the availability objective for each layer network independent of other network layers. But the objective is to meet the end-to-end availability of the network. Even though all transit networks have high network availability except one, this one network with low availability will determine the end-to-end network availability of the network. In general, layer 1 networks have much better network reliability than, for example, layer 3 networks. Therefore, it is not necessary for network planners to improve layer 1 networks without enhancing network availability of layer 3 networks.

A carrier network is comprised of different layers of networks: IP of Layer 3 (L3), ATM, Ethernet and Frame Relay of L2, MPLS of between L2 and L3, and SONET and WDM of L1. Data traffic must travel across the multilayer networks from source to destination. Typically, network planners and engineers designed each layer transit network with the maximum availability independently. While this approach may yield in the optimal design of one-layer transit network, the other transit layer networks may be over-engineered or sub-engineered. To ensure the end-to-end network availability, all transit networks in different layers are required to satisfy one target network availability requirement. This patent application provides a method for designing the end-to-end multilayer network for the optimal overall network system design, by providing one optimal network availability requirement that must be met by all different layer transit networks. This issue is modeled as max-min problem. High-level overview of the method is shown in Fig. 5.18.

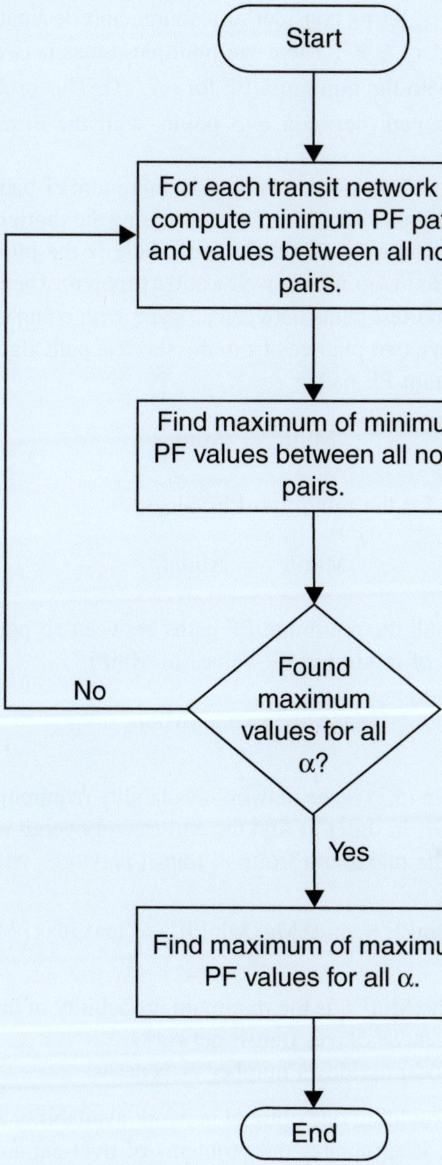

Fig. 5.18 Flowchart of end-to-end multilayer network design method

Consider a layer α network that data traffic must traverse. Any network can be considered as a set of nodes and edges (or links) in graph theory. So we define a layer α network as G^{α}: $G^{\alpha} = (V^{\alpha}, E^{\alpha})$, where V^{α} is a set of nodes, and E^{α} is a set of links for layer α network.

Given $G^\alpha = (V^\alpha, E^\alpha)$, let us consider any source and destination node pair (v_i^α, v_j^α), where $v_i^\alpha \in V^\alpha$ and $v_j^\alpha \in V^\alpha$. There are multiple routes between v_i^α and v_j^α. We need to find the path with the minimum PF for (v_i^α, v_j^α). This problem is equivalent to finding the shortest path between two points with the distance metric is the probability of failure.

For our problem, we are interested in finding the minimum PF paths between all pairs of nodes in G^α. Again, this problem is the same as finding the shortest paths between all pairs of nodes in G^α, where the distance metric here is the probability of failure. Finding the shortest paths in a graph is a well-known problem. There are some efficient algorithms of finding shortest paths between all pairs with complexity of $o(n^3)$.

For G^α, then, we have two matrices from the shortest path algorithm. The matrix is the set of the minimum PF paths:

$$\mathrm{Min}P^\alpha = [\mathrm{Min}p_{ij}^\alpha] \tag{5.1}$$

The other matrix is for the minimum PF values:

$$\mathrm{MinPF}^\alpha = [\mathrm{Min}pf_{ij}^\alpha] \tag{5.2}$$

Once we computed all the minimum PF paths between all pairs of nodes in G^α, we take the maximum of minimum PF values in MinPF^α.

$$\mathrm{MaxMin}^\alpha = \max_{ij}(\mathrm{Min}p_{ij}^\alpha) \tag{5.3}$$

The value found from (5.3) is the network availability requirement for a particular layer transit network, G^α. In order to find the end-to-end overall multilayer network requirement, we take the maximum from all transit networks with different layers:

$$\mathrm{MaxMaxMinPF} = \max_{\alpha}(\mathrm{MaxMinPF}^\alpha) = \max_{\alpha}\max_{ij}\left(\mathrm{Min}p_{ij}^\alpha\right) \tag{5.4}$$

The value of MaxMaxMin*PF* is the minimum probability of failures that must be satisfied by all heterogeneous layer transit networks.

The proposed method provides a number of benefits:

- Cost Savings: By providing one optimal network availability requirement for all layers of networks, it eliminates the problems of over-engineering in typically lower layer networks, which may result in significant cost savings.
- Higher Network Resource Utilization: By avoiding over-engineering in the multilayer networks, network planners can meet the network availability requirements with less resources, thus resulting in higher network utilization.
- Higher Customer Satisfaction: This end-to-end multilayer network design provides the same network availability for networks of all layers, thus avoiding under-engineering in higher layer networks as well as avoiding over-engineering. Customers can potentially expect less network down-times.

5.3 TCP Analytics

The previous sections described the analysis of various network segments by geographical demarcation. In the next two sections, we take a look at protocols and services that span the various network segments end to end. The most important is being a transport protocol—TCP which is a transport service that has many tunable parameters which need measurements in order to understand how to tune to achieve the desired effects.

TCP is probably one of the most important and complex protocols as of data. There are many parameters that need to be measured and tuned for optimal performance. Network analytics at the transport layer is discussed in this section. Again, we present novel and innovative network analytics that the author was involved and patented. The reader is recommended to consult with Verizon Patent and Licensing if any ideas described plan to be commercialized.

Figures 5.19, 5.20, 5.21, and 5.22 show a dissected view of a TCP packet flowing from server to client. TCP was originally designed for networks of equal capacity and all flows sharing the links with equal capacity. However, with wireless networks, all these assumptions are not true and many aspects of TCP need to be

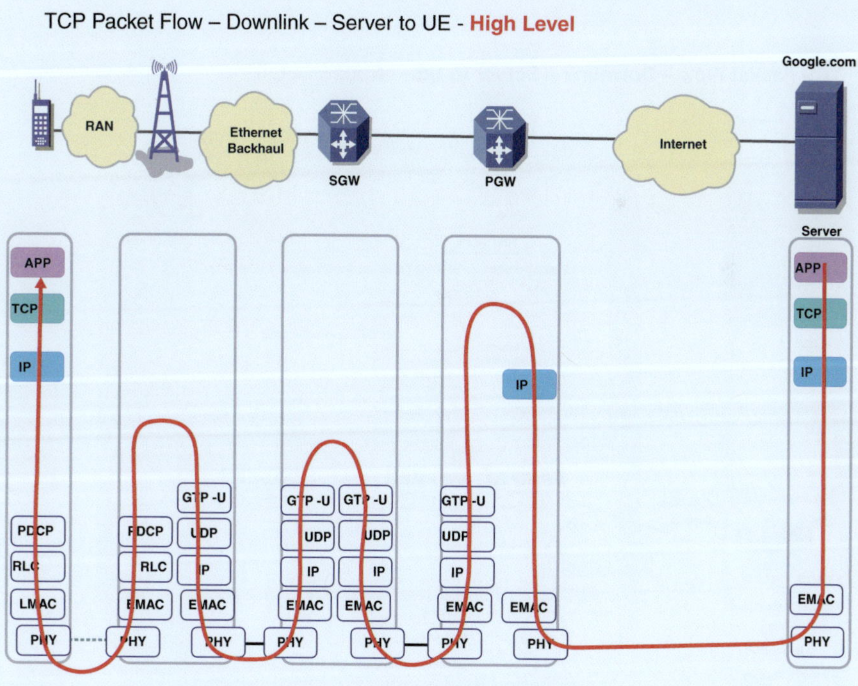

Fig. 5.19 TCP path flow from UE to server end point

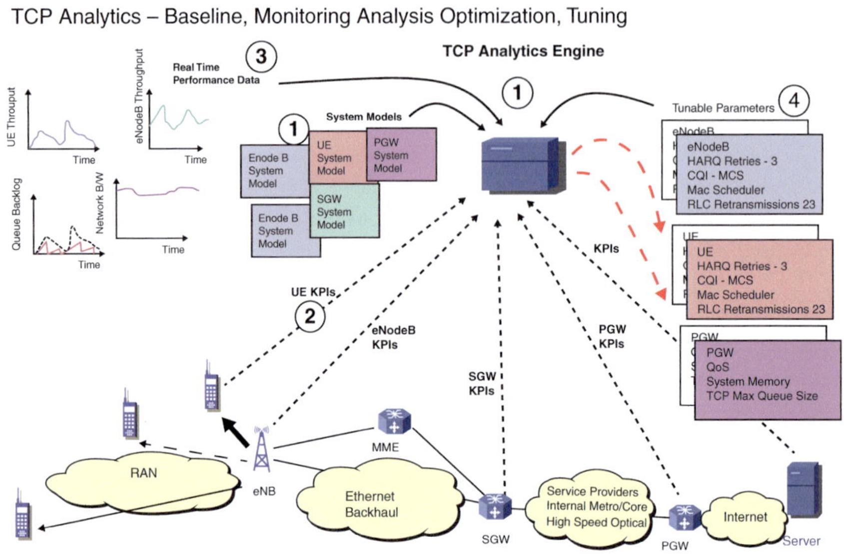

Fig. 5.20 TCP analytics system overview

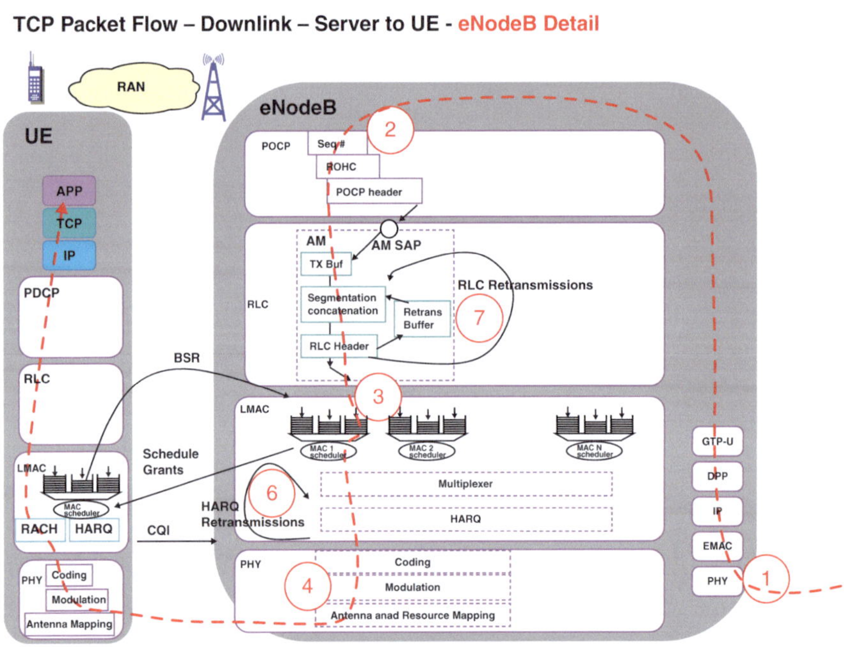

Fig. 5.21 Sources of congestion in EnodeB—bottleneck on downlink

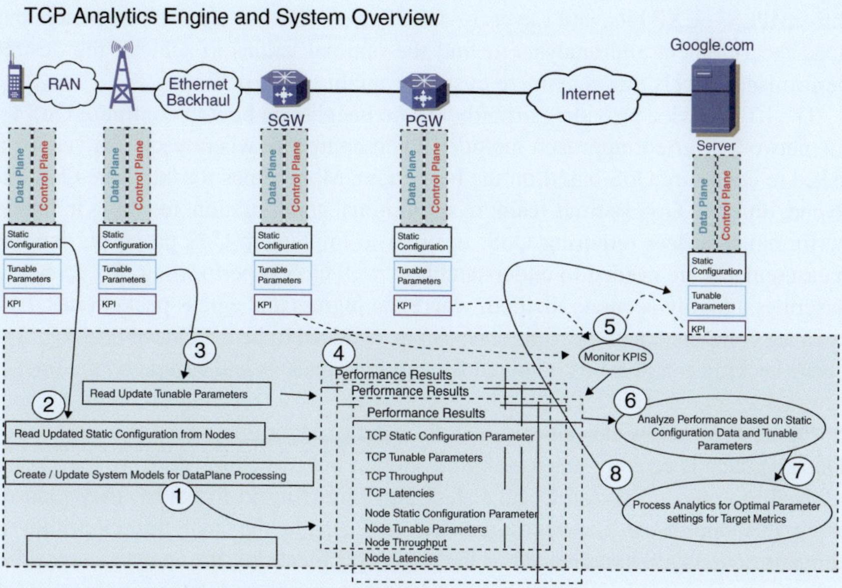

Fig. 5.22 TCP analytics system overview—measurements data flows

adapted for TCP to operate optimally on a wireless network. Example TCP tunable parameters include:

tcp congestion window max
tcp slow start after idle
tcp slow start initial
ssthresh
tcp max buffer size

Some of these parameters can be adjusted dynamically using operating system utilities, for example in Solaris, the utility ndd can be used as

```
#ndd -dev/tcp -set tcp_max_buffer_size 64KB
```

5.4 QoS Analytics

The previous sections described the analysis of various network segments by geographical demarcation and a transport protocol—TCP which is a transport service that has many tunable parameters which need measurements in order to understand how to tune to achieve the desired effects.

In this section, we examine a network service—QoS that spans not only geographical network segments, but also protocols that span Layer 2–802.1pq, Layer

2.5—MPLS—EXP bits, and Layer 3—DSCP. QoS also has tunable parameters that need measurement and analysis to find the optimal values to achieve the desired performance levels that maximize customer quality of experience.

The LTE service provider currently has no insights on how to configure QoS for the network. Current approach includes a silo approach, where each NE vendor is asked to configure QoS based on the best guess. Many times for the same QoS end to end, different QoS settings result in sub-optimal configuration, resulting in poorly performing services requiring QoS. In order to implement QoS properly, ongoing measurements are needed to understand how well QoS is performing, Fix 23 below describes a highly accurate method, where the ingress and egress packet rates, byte rates are computed, and the difference can be deduced to be the Queue backlog. The queue backlog can also be computed by sampling packets and computed the time of egress and ingress (Fig. 5.23).

Figure 5.24 below describes the QoS analytics algorithm again at a high level. This approach is actually patented by the author, but owned by Verizon, readers are advised to contact Verizon Legal Patent and Licensing in the event they plan on using this approach. The measurements will enable the analytics algorithm to compare actual QoS experienced vs the target specifications, the analytics algorithm will then try to minimize the distance between actual and targe across all QoS Levels. Also in consideration is the importance of this distance. For example, it is much more expensive to not meet QCI 1 for voice, than it is for background BE backup data transfers.

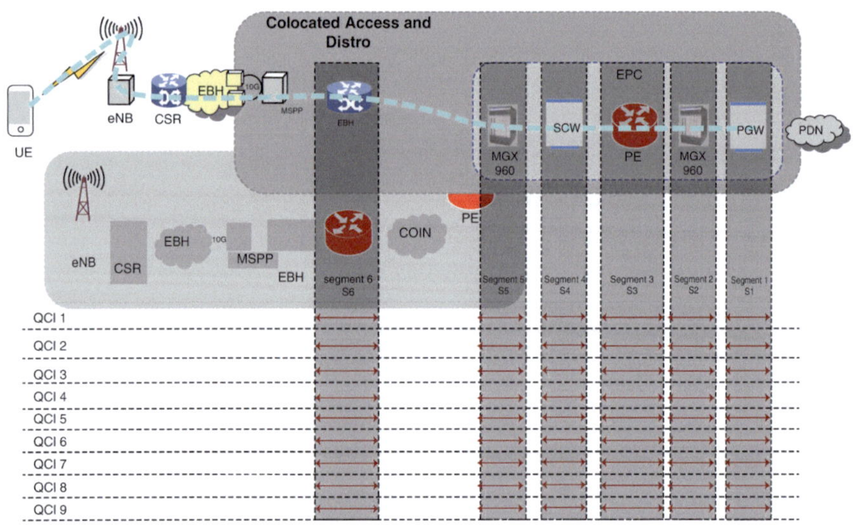

Fig. 5.23 Measuring queue backlog by time stamps and or egress minus ingress byte rate on per QCI per NE basis

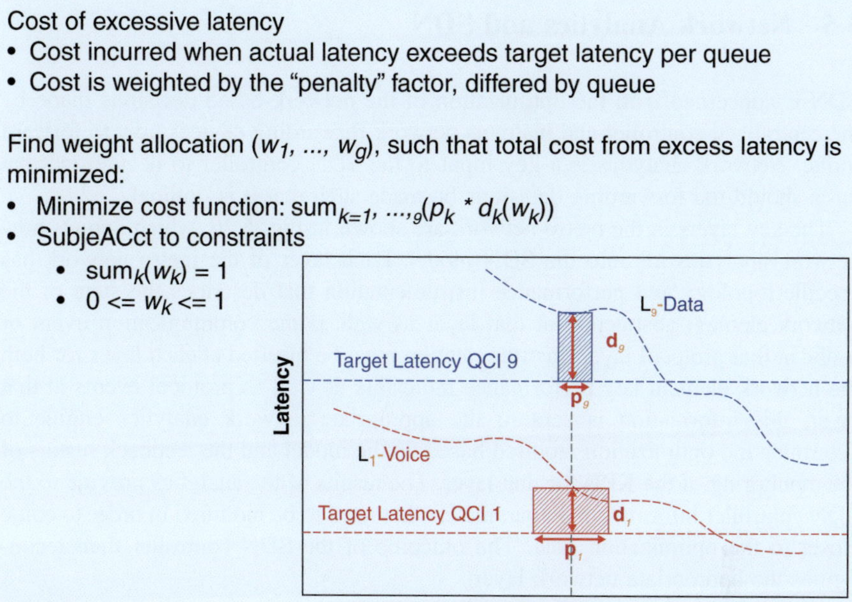

Fig. 5.24 QoS Analytics algorithm—high level—Optimization

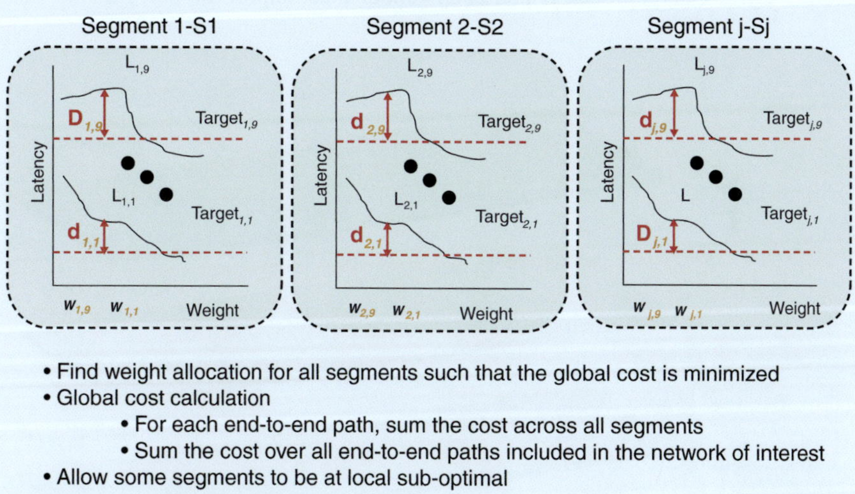

Fig. 5.25 QoS analytics optimization algorithm—global optimization

Figure 5.25 above now extends the algorithm to all NE nodes enable with QoS engines to be optimized, where the notion of distance is extended to all nodes as opposed to a single node.

5.5 Network Analytics and SDN

SDN is concerned with the optimization of the network-based decisions made by the centralized controller and instructs network forwarding devices how to forward traffic. Network analytics is a key input to the SDN controller to decide on what basis should the forwarding decisions be made such that it is optimal.

The key layers in the metro network are shown in Fig. 5.26, which depicts how network analytics fits into the SDN model. Each layer of the metro network has specific topology and performance instrumentation that describes the state of the network element abstraction at that layer as well as the communication event or traffic at that protocol layer. Instrumentation must be inserted at each layer for both the network element key performance indicators as well as protocol events at that layer, this information is sent to the appropriate network analytics engine to determine the optimization required based on the model and the feedback results of the monitoring of the KPIs for that layer. The results of the analytics provide to the SDN controller information on parameters that need to be modified in order to come closer to the optimization goal. The outcome of the SDN controller then reconfigures the appropriate network layer.

In practice, it is very difficult to integrate the data from one layer to another because typically a vendor that has performance monitoring at the optical layer does not have data in a format that can be imported into a tool which is monitoring another layer. One approach is that each vendor must export the performance data

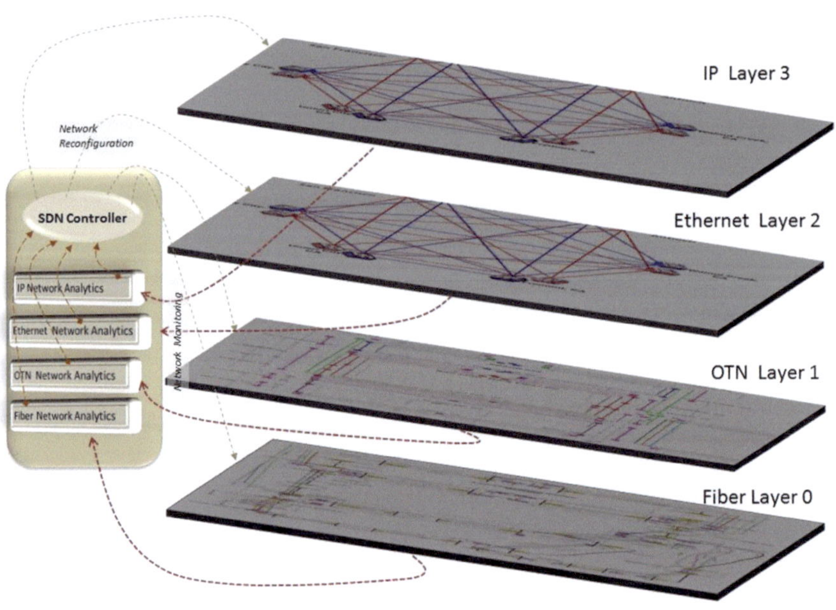

Fig. 5.26 Multilayer network analytics in the SDN model for the metro

in a common normalized and calibrated format that can be ingested by a multilayer network analytics tool which has the network models of each layer and can map the normalized data to process in its algorithms.

5.6 Summary

The LTE wireless has three fundamentally different network architectures, each subtype follows different network analytics metrics and analytic technique yielding specific insights. The Backhaul is usually point-to-point links from mobile switching office to cell sites, which is usually leased from local incumbent service providers, and monitoring takes place at both ends to verify the SLAs promised by the LEC is actually compliant. Network analytics provides insights to verify SLAs are met, and also optimization opportunities to reduce cost and improve performance. The metro network architecture is characterized by optical rings and now more recently hub and spoke designs. The core network is characterized by a mesh which is primarily concerned with reliability, capacity planning, forecasting, and moving massive amounts of aggregated traffic. Network analytics provide insights to optimizations that reduce cost and improve performance.

Chapter 6
Advanced Analytics

In this chapter, we describe limitations of the basic statistical concepts we introduced in Chap. 2, discuss why methods built on basic concepts may result in incorrect conclusions if used without accounting for their assumptions, and provide the reader with advanced theory and tools to address such limitations in order to obtain reliable actionable insights from analyzing LTE performance and fault data.

6.1 Statistical Process Control for E-UTRAN KPIs

In Sect. 2.1, we saw how accessibility metric can benefit greatly from CLT and can be easily fitted into the framework of traditional statistical process control.

The "Dark Side" of Central Limit Theorem

The "dark side" of CLT goes unnoticed in most manufacturing processes. However, when it comes to E-UTRAN, as well as other components of the LTE network, applying Central Limit Theorem makes a lot of things shaky.

6.1.1 Retainability

Two of the three E-UTRAN KPIs, accessibility and retainability, are measures of reliability of connection, one at the E-RAB level and the other at the end user level. We shall first address the retainability, because it is a more obvious example of how CLT can make things fall apart.

Consider that we have an ERAB established. Communication failure counter is incremented as call drops occur, and at the end of the session, we have a full tally of lost calls during the time that the ERAB was running. Applying Eq. (2.2), we have

D. Kakadia et al., *Network Performance and Fault Analytics for LTE Wireless Service Providers*, DOI 10.1007/978-81-322-3721-1_6

the retainability expressed as mean failure rate, which is known to be the inverse of the mean time between failures or MTBF.

Other ERABs are established, and at the end of the life cycle of each ERAB, we have a sampling of MTBFs for the network.

Because each ERAB is independent of others, the Central Limit Theorem dictates that after having run sufficiently many ERABs on a network, we will be have a normally distributed measure of failure rates for this network, with mean equal to the mean of MTBFs and variance equal to the sum of variances of failure rates for each ERAB that we have data from.

To measure the variance for each ERAB session, consider the life cycle of one ERAB. We have a random occurrence of failures. A call can either fail or continue; we model it with binomial distribution and apply the math for Bernoulli trials to measure the mean and standard deviation of this process.

On the surface, it seems that we can apply CLT here as well. All we need is to tally up the average rate of calls made during the ERAB session $\left(\text{callRate}_{\text{ERAB}} = \frac{N_{\text{calls}}}{\text{duration of ERAB session}}\right)$, and we can compute the probability of failure:

$$p = P\{\text{failure}\} = \frac{\text{mean failure rate}}{\text{mean call rate}} = \frac{\text{Retainability}}{\text{callRate}_{\text{ERAB}}} \tag{6.1.1}$$

From here, we can establish the variance of the failures during each ERAB session:

$$\text{variance}_{\text{failures}} = N_{\text{calls}} * p * (1 - p) \tag{6.1.2}$$

Substituting Eq. (6.1.1) for each ERAB session into (6.1.2), we have the full description of the asymptotically normal distribution of failures for the network:

$$\text{mean}(N_{\text{failure}}) = \frac{\sum_{i=1}^{N_{\text{ERAB}}} N_{\text{calls}}[i] * p[i]}{N_{\text{ERAB}}} \tag{6.1.3a}$$

$$\text{variance}(N_{\text{failure}}) = \sum_{i=1}^{N_{\text{ERAB}}} N_{\text{calls}}[i] * p[i] * (1 - p[i]) \tag{6.1.3b}$$

Problems with This Approach

Failure arrival is a memoryless process: If at time t there was a failure, it does not mean that there will be a failure at time $t + \Delta t$, at $\forall(t.\,\Delta t)$. By the same token, if at time t the system (the ERAB) was operational, it does not mean that it will be operational at time $t + \Delta t$, at $\forall(t.\,\Delta t)$.

The "Right" Way to Compute Retainability for the Network
The only memoryless distribution known today is the exponential distribution, which provides the probability of event (failure) occurring at any time *t* given event arrival rate:

$$P\{\text{failure at } T \le t\} = 1 - \mathrm{e}^{-\lambda * t} \tag{6.1.4}$$

Here $\lambda = \text{failure rate} = \frac{1}{\text{MTBF}} = \text{Retainability}$

Equation (6.1.4) makes it possible to build probability distribution of failures at any time t from the start of the ERAB life cycle.

We see that in this hypothetical simulation, 95% calls will experience a loss between 12 and 19 s, and after 30 s, over 98% calls will experience a loss. This provides a useful piece of information for designing a fault-tolerant network.

In Fig. 6.1, the bundle of lines describing the dynamics of call loss probability has been produced using the simulation code below:

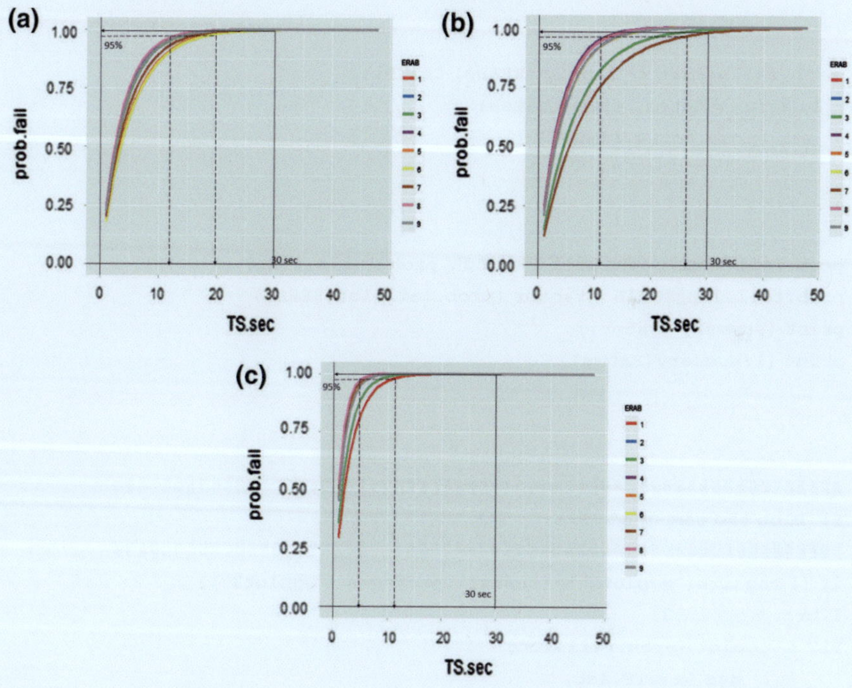

Fig. 6.1 a–c Probability dynamics of call failure at different failure rates (simulated data)

```
##############################################################
## Simulate ERAB call failure process
##############################################################
Ch2.Prob.Fail <- function (erab, Rates, time){
 return (data.frame (
  ERAB = erab,
  Retainability = Rates[erab],
  TS.sec = time,
  prob.fail = (1 - exp (-Rates[erab] * time))
 ))
}
Ch2.Reliability.Analysis <- function (ERABs, Rates, times){



 #  indx =5; Rates = Rates.list[[indx]]



  prob.fails.list <- lapply (ERABs, Ch2.Prob.
Fail, Rates = Rates, time = times)
 #  str (prob.fails.list)



 prob.fail.long <- do.call ('rbind', prob.fails.list)
 prob.fail.long$ERAB = factor (prob.fail.long$ERAB)
 print (summary (Rates))
 print (1/summary(Rates))



 ##############################################################
 ## Plot the computed CDFs
 ##############################################################
 if (! require("ggplot2")){install.packages ("ggplot2")}
 library(ggplot2)
 PLT <- ggplot (prob.fail.long,
          aes (x = TS.sec,
             y = prob.fail,
             colour = ERAB,
             group = ERAB
          )
 )
```

```
  print (
   PLT +
    geom_hline (aes (yintersept=0.95)) +
    geom_vline (aes(xintersept=40)) +
    scale_colour_brewer (palette = "Set1") +
    geom_line(size = 1.5)
  )


  return (prob.fail.long)
}
ERABs <- 1:9
MINS <- runif (10, 0.05, 0.2)
MAXS = runif (10 , 0.27, 0.39)
Rates.list <- lapply (1:10, function (ii, n, MINS, MAXS){
  runif (n, min (c (MINS[ii], MAXS[ii])), max (c (MINS[ii], MAXS[ii])))
}, n = length (ERABs), MINS = MINS, MAXS = MAXS
)
str (Rates.list)
times <- 1:(24*2) #Compute probabilities of failure in 2 days
prob.fail.long.list <-
  lapply (1:length (MINS), function (indx, MINS, MAXS, ERABs, Rates.list,
times){


  #  indx <- 1
  cat (indx, "From", MINS[indx], "To", MAXS[indx], "\n")
  retDF <- Ch2.Reliability.Analysis(ERABs = ERABs,
                        Rates = Rates.list[[indx]],
                        times = times
  )



  retDF
  }, MINS = MINS, MAXS = MAXS, ERABs = ERABs, Rates.list = Rates.list,
times = times
)
```

Analysis similar to that in Fig. 6.1 can be done on real data (by replacing the "Rates" line with real measured data from the E-UTRAN's ERABs), giving important actionable insights recommending optimal ERAB operation runtimes.

Retainability and SPC

As we have demonstrated, direct application of SPC to retainability may be misleading. However, Fig. 6.1 provides an interesting new twist on the old problem. In that example, we saw that by 45 s, 98% of the ERABs will have failed. This means that, based on the nine ERABs that we had in operation, we have 98% confidence that this E-UTRAN will fail in 45 s.

On the other hand, the time interval corresponding to 95% call loss (between 22 and 32 s) already provides an interval of time values corresponding to 95% confidence that an ERAB will fail.

If in the next (after this hypothetical example) run of 9 ERABs, we collected another set of Retainability data, the 95% range of times to fail and the 30-s probability of failure would be different.

Here, the 95% failure time range is 12–27 s, and only 96% of all ERABs will experience a failure at 30 s.

In a still next set of ERABs, we have a different picture that 95% of calls fail within 4–12 s, and none of the calls survived the 30-s threshold.

What do we see from Fig. 6.1a–c? The quality of service (Retainability) is highly volatile: the 95% failure range start point and width, as well as the 30-min failure probability, are changing from one set of ERABs to the next. This is the kind of variability that we want to show on SPC charts.

The X-minute probability of failure (in our examples, $X = 30$) is a binary variable: At X minutes from the ERAB establishment, the calls may either fail or succeed. For such metrics, we have the already familiar np- and p-charts, such as in Figs. 6.1 and 6.2.

While the 95% failure range is more meaningful than X-minute failure probability (it reflects both the movement and the uncertainty of failure predictions), it is still not normally distributed, and its interpretation is more complicated than the X-minute failure probability. Interested readers are welcomed to explore the topic further elsewhere in the literature.

It can be demonstrated that for a normal distribution, the "3-sigma" interval corresponds to 99.7% confidence. This should enable us to formulate equivalent SPC confidence intervals (aka control limits) for any metric.

6.1.2 The KPIs: UL and DL Throughputs

As we shall describe in Sect. 3, throughput in a queueing system usually can be described with high degree of accuracy as a Poisson distribution. One of the nice features of the Poisson distribution is that at high rates (throughputs), it becomes approximately normal, the difference being on the tails. This gives us a chance to use traditional SPC, at least for high throughput values.

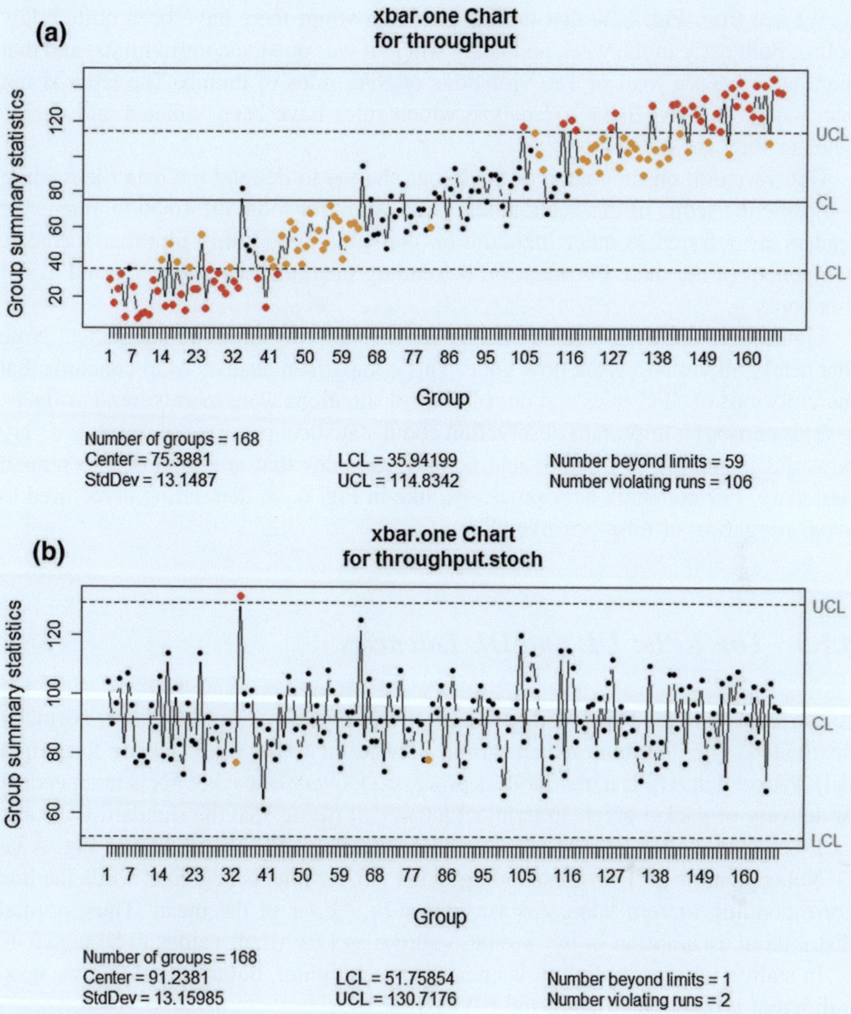

Fig. 6.2 **a** and X-bar SPC chart for throughput. **b** Detrended throughput data: stochastic component

As we are not dealing with probabilities anymore, a different chart will have to be used for throughput SPC. A number of SPC charts are widely used for continuous variables. Below is an example of an X-bar chart produce by using the following three lines of R code:

```
if (! require ("qcc")){install.packages ("qcc")}
library (qcc)
xChart <- qcc (throughput, sizes = 10, type = "xbar.one")
```

We see from Fig. 6.2a that throughput is growing; there have been quite a few points, both early in the week and lately where it was outside control limits, and that there have been a total of 150 violations of SPC rules of thumb. The latter is not necessarily bad: We have to analyze which rules have been violated and decide whether they are critical.

One variation on the theme of the X-bar chart is to detrend the data (detrending is outside the scope of this section; see Regression; for more information, interested readers are referred to other literature on the subject) and only plot the stochastic component of the data. For ideas on detrending methods, see Sects. 2.4 and 2.5 of this book.

Figure 6.2b shows the same data as above, but with the trend subtracted. Note that nearly all violations are now gone. This comparison enables us to conclude that the violations of SPC rules and out-of-control situations were merely trend artifacts.

This carries an important observation about statistical process control: It is a very powerful technique to monitor and control variables that are supposed to remain stationary. For non-stationary processes, like in Fig. 6.2a, detrending is required to avoid triggering of false-positive alarms.

6.1.3 The KPIs: UL and DL Latencies

As we showed in the Retainability section, times between events are never normally distributed. The distributions tend to be skewed far to the right (have a long right tail). Packet delivery is a memoryless process: Delivery of packet #N is independent of delivery of packet #N-1. In terms of latency, it means that the standard UCL and LCL calculations for latency will lead to misleading SPC charts, like in Fig. 6.3a.

Notice that the LCL, corresponding to the $-3 * \sigma$ line, is negative, while the line corresponding to zero latency is very close to $-1 * \sigma$ of the mean. Thus, normal distribution assumption in this scenario allows $\sim 17\%$ of all values to be negative.

In reality, the distribution of latencies is much tighter, but to reveal that we need a different type of SPC chart, the EWMA chart.

Note that the EWMA chart in Fig. 6.3b has two sets of data points—the "+" corresponds to observed data, whereas the "·" and the solid line correspond to the points fitted into the observed data by the EWMA technique. Details of EWMA are outside the scope of this book.

Also note that the points marked red are only on the EWMA line, though there are many observed data points above the UCL line. This is happening because the UCL and LCL are built based on the EWMA data. In other words, it is still an approximation, and the $-3 * \sigma$ lines are still symmetrical around the mean, failing to reflect the nature of the distribution. However, "all models are wrong, but some are useful" (George E.P. Box), and this is certainly a more useful model than the X-bar chart.

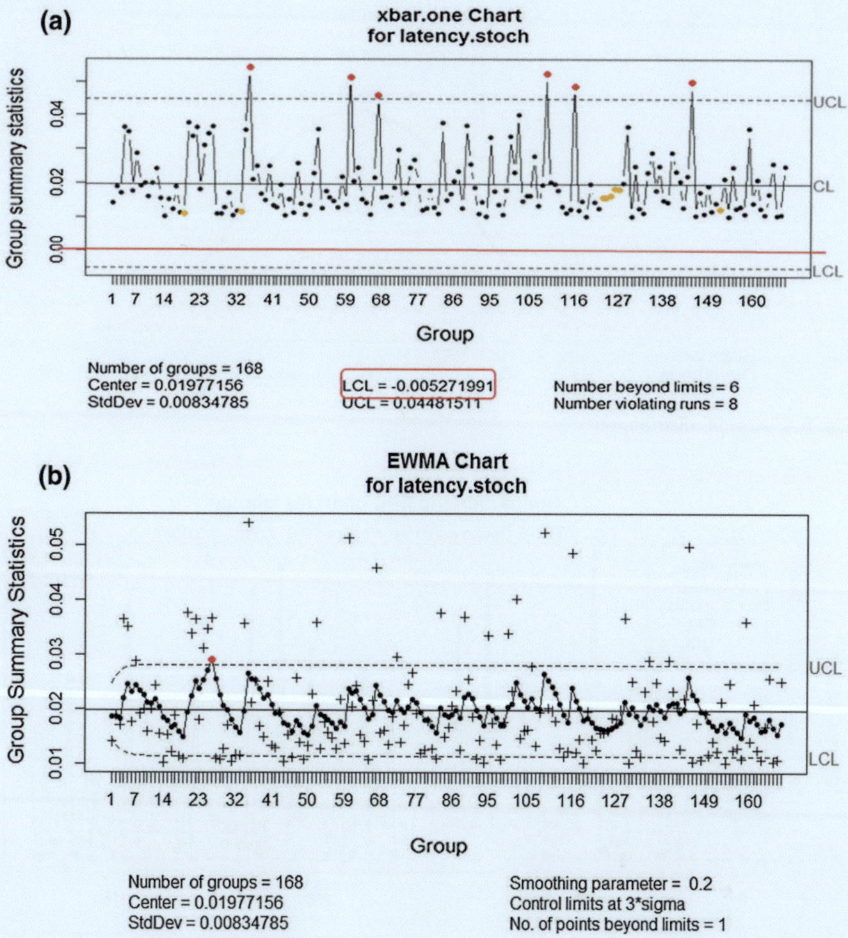

Fig. 6.3 **a** X-bar SPC latency still tries to fit a normal distribution around latency. **b** SPC for Gaussian (normal) latency

6.1.4 Nonparametric SPC—A Cleaner Way to Show the UCL and LCL of the Non-normal Variable

Let us consider a classic bell curve (density plot) of a normal distribution.

We can see for Fig. 6.4 that $\mu \pm 3 * \sigma$ range (where $\mu = \text{mean}, \sigma = \text{standard deviation}$) corresponds to 99.74% of all data. In other words, the 0.13th percentile line will correspond to $\text{LCL} = \mu - 3 * \sigma$, whereas the 99.87th percentile will be the line for $\text{UCL} = \mu + 3 * \sigma$.

The following lines of R code produce the percentile SPC chart in Fig. 6.5:

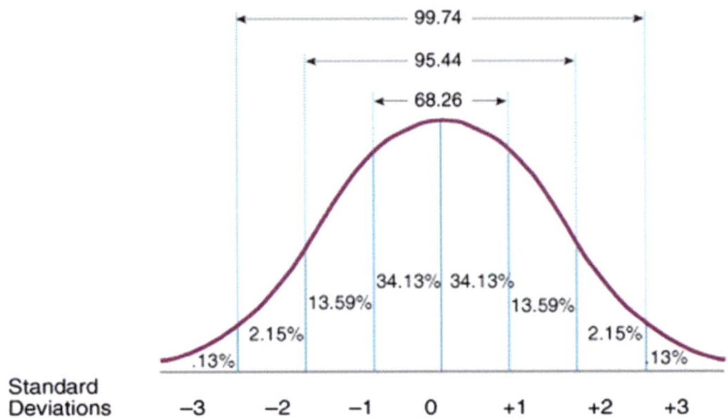

Fig. 6.4 Normal distribution

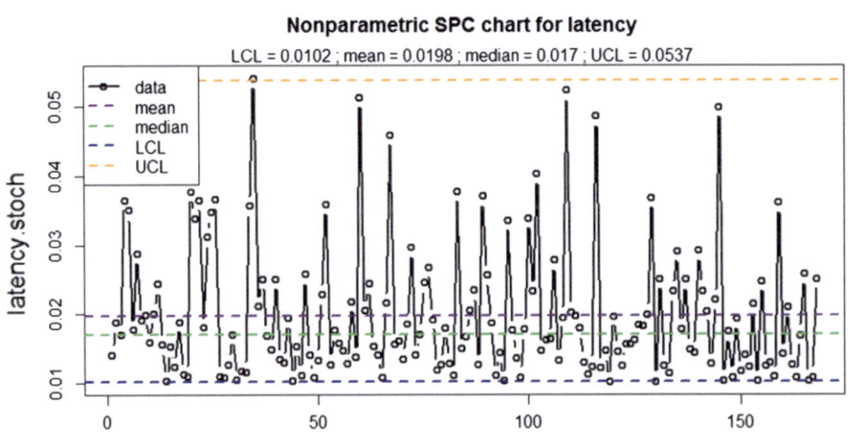

Fig. 6.5 Nonparametric SPC chart for latency data

```
##############################################################
## Plot the nonparametric SPC chart
##############################################################
plot.ts (latency.stoch, type = 'b', lwd = 2,
     main = "Nonparametric SPC chart for latency")
abline (h = mean (latency.stoch), col = "purple", lty = 2, lwd = 2)
abline (h = median (latency.stoch), col = "green", lty = 2, lwd = 2)
abline (h = quantile(latency.stoch, prob = c (0.0013, 0.9987)),
     col = c ("blue", "orange"), lty = 2, lwd = 2)
legend (x = "topleft",
```

```
    pch = c (1, NA, NA, NA, NA),
    lty = c (1, 2, 2, 2, 2),
    lwd = c (2, 2, 2, 2, 2),
    col = c ("black", "purple", "green", "blue", "orange"),
    legend = c ("data", "mean", "median", "LCL", "UCL")
    )
mtext (
 paste ("LCL =",format (x = quantile (latency.stoch, 0.0013), digits = 3),
     "; mean =",format (x = mean (latency.stoch), digits = 3),
     "; median =",format (x = median (latency.stoch), digits = 3),
     "; UCL =",format (x = quantile (latency.stoch, 0.9987), digits = 3)
  )
 )
```

Figure 6.5 shows most of the information we are looking for: the data, UCL, LCL, and two CL lines (mean and median). Because implementation of the SPC rules is not part of an existing R package, it is up to the analyst to develop code that will implement them.

Naturally, the same code can be implemented for any data, regardless of the distribution—retainability, throughput, accessibility, and other KPIs, as well as their linear and nonlinear combinations.

6.1.5 The KPIs: Availability

Availability is mathematically same as accessibility, but for a continuous variable, it is a measure of probability of the E-UTRAN (or ERAB, or any other element) being available at any moment in time. We can apply the p-chart SPC for monitoring this KPI.

6.1.6 The KPIs: Mobility

Mobility is a product of two probabilities of success—handoff preparation and handoff execution. The product of two normal (or binary) variables is not normal, even asymptotically. Therefore, we can use the same nonparametric approach as we have formulated for latencies: compute and show on the plot the 0.13th and the 0.9987th percentiles, mean, and median.

6.2 Real-World Outliers

"It may be normal, darling; but I'd rather be natural."

(Truman Capote, *Breakfast at Tiffany's*)

6.2.1 Introduction

In Sect. 2.2, we discussed outliers and the need to differentiate between outliers and SLO violations, as well as the basic method of detecting outliers by superimposing the $\pm 3\sigma$ band on the data and detecting the violators of the "3-sigma rule."

Consider now a generic gamma distribution. Gamma distributions are often observed in the IT world in general. By properly selecting its parameters, we can use a gamma distribution to approximate the Poisson, exponential, Erlang distributions, as well as any distribution in between, because they are all special cases of the gamma.

An example of such distribution is represented by the histogram in Fig. 6.6.

Finally, the cumulative distribution function will be given by Fig. 6.7.

The tuple ($Q1$, $Q2$, μ, $Q3$, σ) fully (and even redundantly) characterizes a distribution. Some programming languages have evolved to automatically write down the summary of any distribution. For example, in R, it will be:

```
summary (mydata)
```

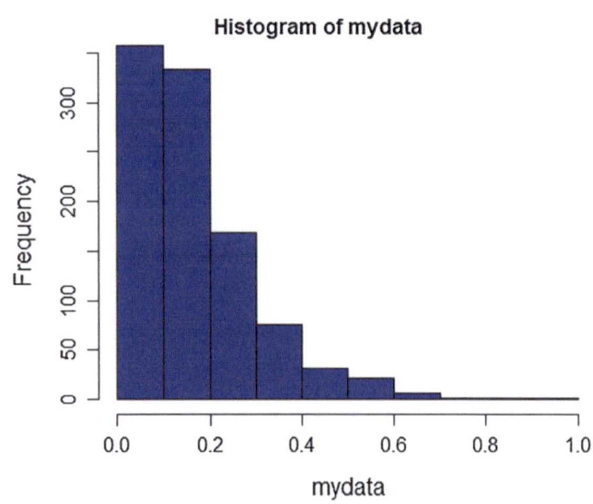

Fig. 6.6 Histogram of a gamma distribution

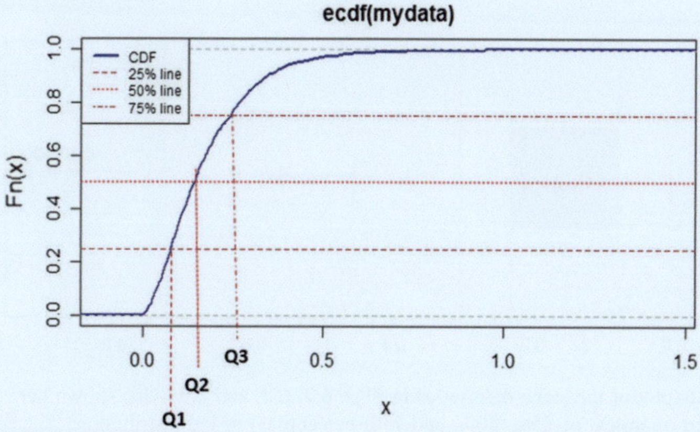

Fig. 6.7 Cumulative density function (CDF) of the same data set as in Figs. 2.4 and 2.5. In addition, shown here are the 1st, 2nd, and 3rd quartiles (or 25th, 50th, and 75th percentiles) of the distribution

The summary(data) command will return a vector containing the main elements of the distribution:

Min.	1st Qu.	Median	Mean	3rd Qu.	Max.
0.001934	0.076310	0.133400	0.169200	0.227900	0.917100

However, this array is missing two important items: (i) standard deviation and (ii) an indication of whether there are any outliers in the data.

In the 1950s, John Tukey proposed a clever distribution agnostic method for the identification of outliers in any data:

1. Compute the inter-quartile range as IQR $= Q3 - Q1$.
2. Compute the "whiskers":
   a. LW $= Q1 - 1.5 *$ IQR;
   b. HW $= Q3 + 1.5 *$ IQR.
3. Any points outside the whiskers are outliers.

This method later became known as the boxplot method. For our data, a boxplot is presented in Fig. 6.8.

The short vertical lines in Fig. 6.8 correspond to the bounding numbers of the whiskers. The data points (small circles) outside the whiskers are outliers.

For reasons unknown to us, R does not have a built-in method to directly identify outliers. However, it is easy enough to write a function that will do just that:

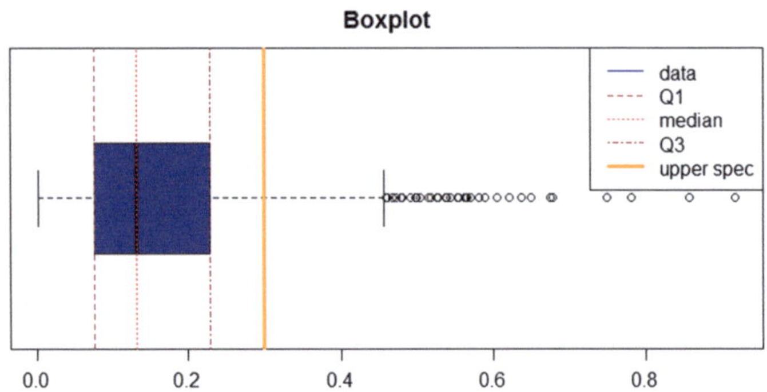

Fig. 6.8 Boxplot of the same data set as in Figs. 6.7, 6.8, and 6.9. Also shown here are the 1st, 2nd, and 3rd quartiles (or 25th, 50th, and 75th percentiles) of the distribution

```
GetOutlierIndices = function(X, Upper = TRUE, Lower = TRUE){
 # Upper = TRUE; Lower = TRUE ### for debugging
 qrtls = quantile(X, probs = c (0.25, 0.75))
 upper = qrtls[2] + 1.5 * (qrtls[2] - qrtls[1])
 lower = qrtls[1] - 1.5 * (qrtls[2] - qrtls[1])
 UpperOLindices = which (Upper & X > upper)
 LowerOLindices = which (Lower & X < lower)
 return (c(LowerOLindices, UpperOLindices))
}
```

This function will return the indices for upper and lower outliers together. Because for our domain, we are mostly interested in high outliers (we will talk about use cases for lower outliers later in this book), we can specify `Lower = FALSE` in the arguments when calling the function.

Note that this function returns indices of outliers in vector *X* (the **mydata** object). To find the values of outliers, all we need is five lines of code:

```
myOLs <- GetOutlierIndices(X = mydata, Upper = T, Lower = F)
myOL.values <- NULL
if (length (myOLs) > ){
 myOL.values <- mydata[myOLs]
}
```

6.2.2 Advantages of Tukey's Method

1. It is distribution agnostic.
2. It directly relies on the probabilistic foundations of statistics.
3. It is mathematically simple.

6.2.3 Disadvantages

1. There is no direct correlation between outliers and confidence levels.

6.2.4 Practical Use Cases

Decision Support
If the **mydata** object contains latencies (packet delays), we can check whether the cases where the packet delay budget was exceeded were outliers or were part of the distribution.

```
#################################################################
## Identify measurements where packet delay budget was exceeded
#################################################################
packet.delay.budget <- 0.3 ## 300 ms


exceeded.indcs <- which (mydata > packet.delay.budget)
any.outliers.TF <- NULL
all.outliers.TF <- NULL
if (length (exceeded.indcs) > 0){
 ## Are any of them outliers?
 any.outliers.TF <- any (myOLs %in% exceeded.indcs)
 ## Have all outliers exceeded the budget?
 all.outliers.TF <- all (myOLs %in% exceeded.indcs)
}
any.outliers.TF
all.outliers.TF
```

This will return, in our case, **TRUE** and **TRUE**.

The instances that we should be concerned about as first priority are non-outliers that have exceeded the packet delay budget. In the example that we looked at, we had 140 data points (out of 1000) outside the delay budget of 300 ms. Of these, 99 were not outliers, and 41 were outliers. If these were real data from an actual EPC, we would need to seriously look into this specific instance of EPC, for two reasons: (i) Only 86% of the data is within the packet delay budget; (ii) 41% of the out-of-spec data is outliers. This process is both out of specifications and out of control, and either the loading has to be reduced or the bandwidth has to be increased.

Deep-Dive Analysis: Beyond Outliers

We shall present here an example of deep-dive analysis applied to a scenario where packet delays were sampled with 5-min intervals for 100 UEs over the course of 12 morning hours.

That generated a sampling of 144 samples for each of the 100 UEs. Each UE had its own bandwidth usage patterns, and the packet drop ratios stayed within the specifications. However, packet delays were not always within specifications.

We see from Fig. 6.9 that all UEs had outlier samples, and most UEs have exceeded the 300-ms specification limit at least once. There does not seem to be any relationship between delay and the end user ID. We shall see whether we can still infer any useful information about the data from the samples that we have.

The data have been collected into a tabular form (Table 6.1).

Plotting the delay data versus the time stamp (Fig. 6.10) shows us two important things:

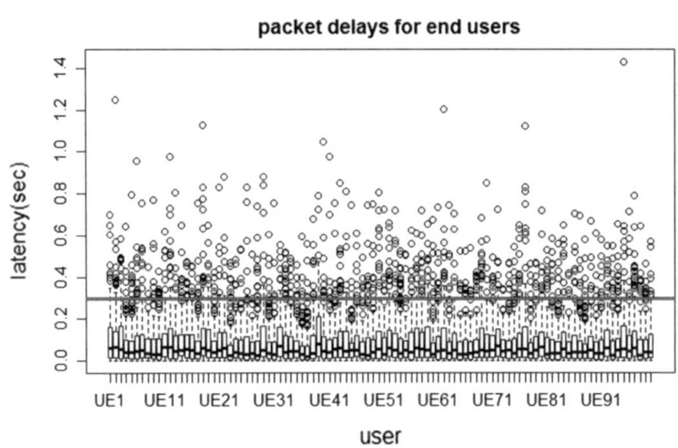

Fig. 6.9 Packet delay samples for 100 UEs

Table 6.1 Delay samples for UE1 for the first 90 min

UE	Delay	Time stamp	Hour	Busy
UE1	0.159492	1	6	QUIET
UE1	0.000949	2	6	QUIET
UE1	0.241452	3	6	QUIET
UE1	0.006681	4	6	QUIET
UE1	0.103489	5	6	QUIET
UE1	0.026324	6	6	QUIET
UE1	0.005141	7	6	QUIET
UE1	0.028138	8	6	QUIET
UE1	0.258149	9	6	QUIET
UE1	0.166097	10	6	QUIET
UE1	0.149063	11	6	QUIET
UE1	0.024867	12	7	QUIET
UE1	0.012372	13	7	QUIET
UE1	0.128344	14	7	QUIET
UE1	0.426067	15	7	QUIET
UE1	0.100197	16	7	QUIET
UE1	0.030671	17	7	QUIET
UE1	0.187387	18	7	QUIET
UE1	0.041188	19	7	QUIET
UE1	0.101391	20	7	QUIET
UE1	0.000157	21	7	QUIET
UE1	0.257052	22	7	QUIET
UE1	0.072642	23	7	QUIET
UE1	0.00335	24	8	QUIET
UE1	0.013906	25	8	QUIET
UE1	0.054306	26	8	QUIET
UE1	0.603689	27	8	QUIET
UE1	0.011645	28	8	QUIET
UE1	0.050361	29	8	QUIET
UE1	0.019038	30	8	QUIET
UE1	0.012697	31	8	QUIET
UE1	0.162575	32	8	QUIET
UE1	0.017876	33	8	QUIET
UE1	0.070851	34	8	QUIET
UE1	0.035307	35	8	QUIET
UE1	0.002934	36	9	QUIET
UE1	0.00263	37	9	QUIET
…	…	…	…	…
UE1	0.000585	49	10	BUSY
UE1	0.002656	50	10	BUSY

(continued)

Table 6.1 (continued)

UE	Delay	Time stamp	Hour	Busy
UE1	0.343759	51	10	BUSY
UE1	0.541089	52	10	BUSY
UE1	0.149759	53	10	BUSY
UE1	0.417197	54	10	BUSY
UE1	0.360537	55	10	BUSY
UE1	0.256445	56	10	BUSY
UE1	0.030083	57	10	BUSY
UE1	0.128365	58	10	BUSY
UE1	0.002728	59	10	BUSY
UE1	0.232837	60	11	BUSY

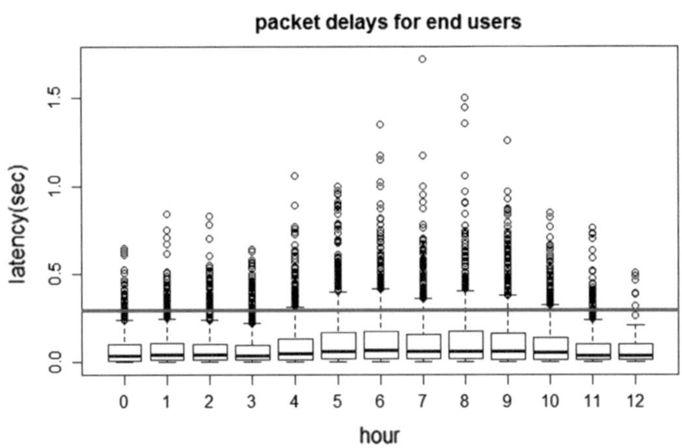

Fig. 6.10 Hourly distributions of packet delays

1. There seems to be a nonlinear correlation between the delay and the hour.
2. The worst delay is in the busy hours (between 10:00 and 14:00). This is a hypothetical scenario, in which we want to demonstrate the method without focusing on diurnal patterns typical of LTE user element busy versus quiet hour patterns.

We see that the distribution of outliers by hour is not very interesting, while outlier distribution by UE (Fig. 6.9) is not very revealing either. However, an interesting, and important, insight coming from these data is extracted and analyzed below. We will see whether any UEs have more outliers than others.

6.2.5 Aggregating Data

Often, like in this example, we only have 12 data points for each (UE, hour) pair. This is usually not enough to detect outliers in each pair. However, it is possible to aggregate data into larger bins (e.g., am and pm; morning, afternoon, and evening; busy hours and quiet hours). This will give us 3–6 times as many data points, which is usually sufficient to perform a meaningful analysis. If we find a bin within which the system is misbehaving worse than others, then we can dive deeper into this bin to extract more useful information from the data.

Let us consider aggregation by busy hour.

We will designate hours between 04:00 and 10:00 as busy, set the remaining hours as quiet by adding a "busy" flag to the data set, and plot delay data as a function of this flag. This immediately confirms the hypothesis that busy hours see more out-of-spec delays (Fig. 6.11).

This plot immediately discloses a lot of important information:

1. That data is, in both cases, right-skewed (has a long tail to the right)
2. That during the quiet hours, only outliers are out of specs
3. That busy hour distribution is both wider and higher than during the quiet hours

Now we can compare the sets of outliers during the busy and the quiet hours. We know that their values are higher during busy hours, but we are looking for overlaps in UE names, to see whether there are any hot spots, where there were outliers during both busy and quiet hours.

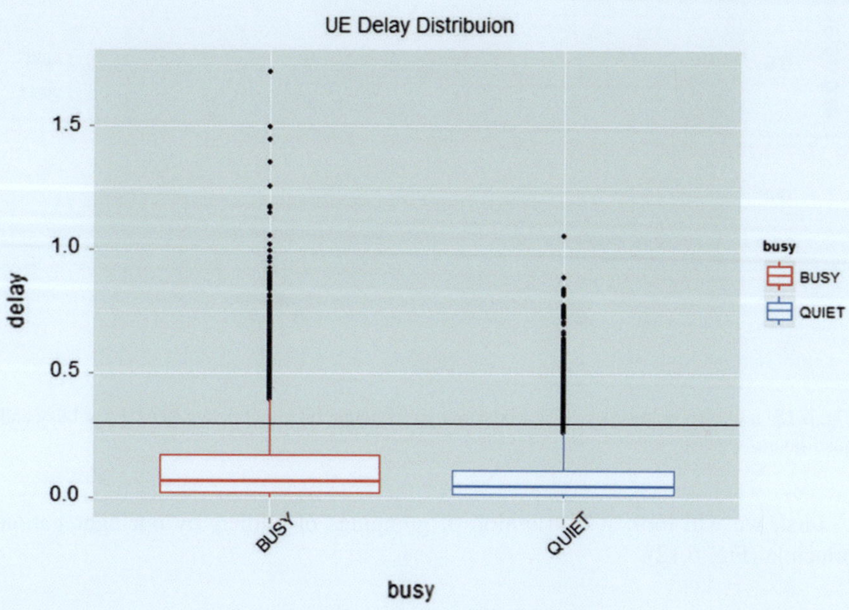

Fig. 6.11 Delay distribution aggregation by busy versus quiet hours

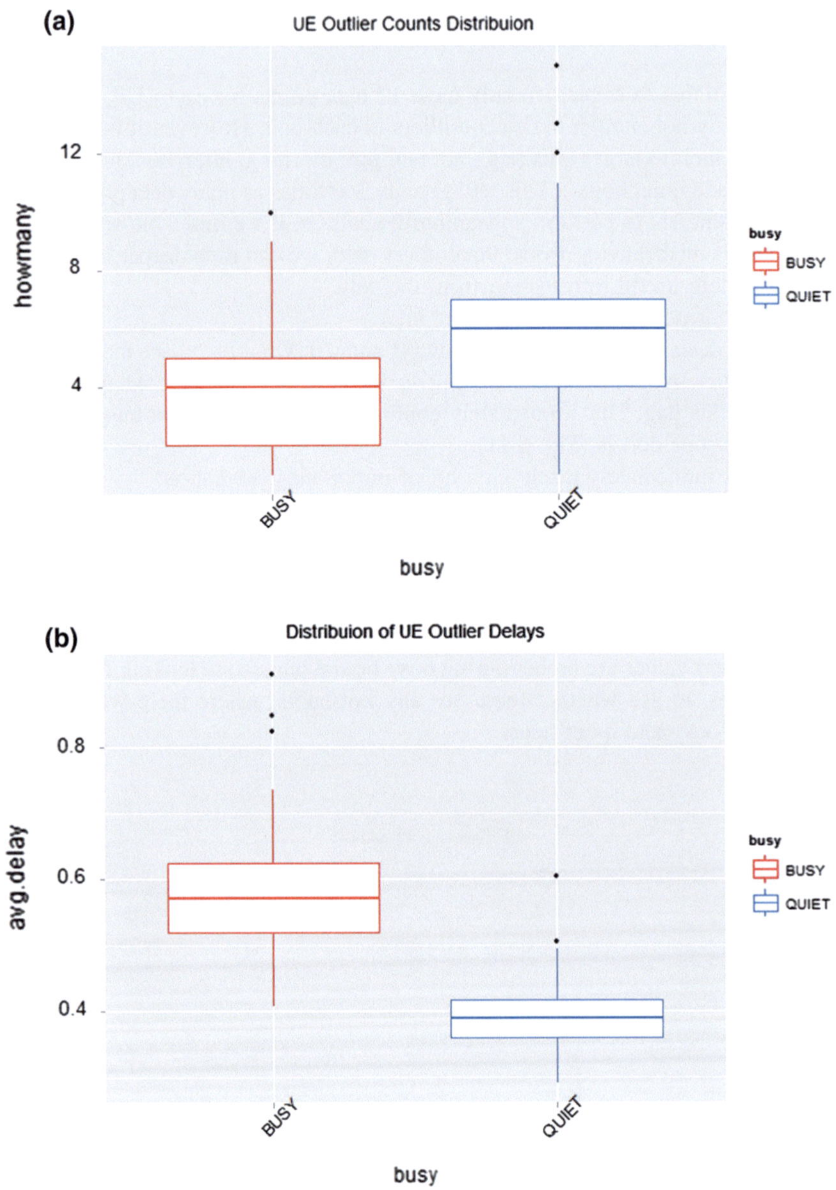

Fig. 6.12 **a** Outlier counts per UE for busy and quiet hours. **b** Outlier delays per UE for busy and quiet hours

First, we will look at distribution of groupings of outliers by our aggregation principle (Fig. 6.12).

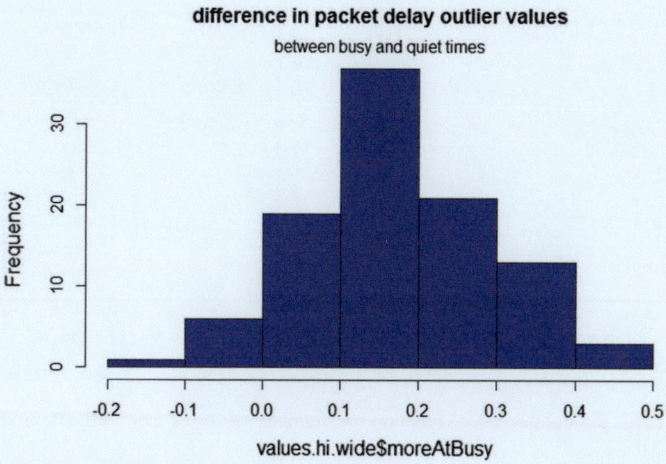

Fig. 6.13 Difference between outlier delays per UE during busy versus quiet hours

We see that there are more outliers during the quiet hours, but their magnitude is greater during the busy hours. To quantify this qualitative statement, we look at the average (mean) values of outliers: looking at the sums would be misleading in this case: first, these are delays per UE, and a sum of all delays does not make any sense, and second, we need to take out the weighing effect of the outlier counts.

Further examination of the data fails to find any hot spots or any correlation with UE ID (more on correlations and regressions in the next chapter).

The histogram of the differences, for each UE, between the packet delays during busy and quiet hours (Fig. 6.13) reveals that

1. Central Limit Theorem works even on outliers of a very right-skewed distribution
2. On average, delays during busy hours are longer
3. The busy hour extension of packet delays varies with mean = 176 ms and standard deviation = 116 ms

Reader is encouraged to perform similar analysis with different cuts of the data (e.g., am/pm; morning/afternoon/evening; location—if such data is available).

6.3 Queueing and LTE Data

6.3.1 User Throughput

As outlined in [SCHW2013], user throughput lines follow a profile very similar to profiles predicted by USL (Fig. 6.14).

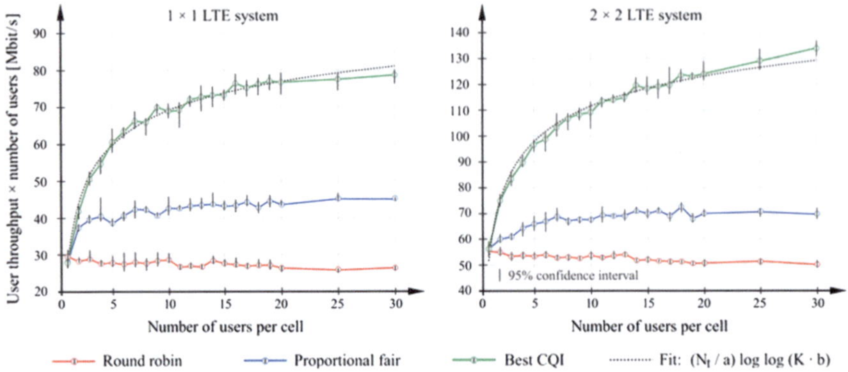

Fig. 6.14 Total throughput as a function of number of users per cell (Reproduced from [SCHW2013])

The standard model used by Schwartz et al. for such systems is the so-called $N_t \log \log k$ rule, the theory for which was developed by Heath et al. (2001):

$$\text{Throughput} = \frac{N_t}{a} * \log(\log(k * b)) \tag{6.3.1}$$

This rule, however, does not account for the dynamics of the transmission system, as the authors explicitly assume that "the bandwidth of the signal is much less than the coherence time of the channel." Therefore, caution must be exercised when using the $N_t \log \log k$ rule approach.

6.3.2 *Effects of Data Transformations on Distributions*

Logarithmic transformation is one of the best-known transformations applied to right-skewed (long right tail) to make them "look normal." They are very efficient.

We see from Fig. 6.15 that even a single logarithmic transformation has turned a right-skewed distribution into a much more symmetrical one.

It is reasonable to assume that a double log transform of a normal distribution will be skewed to the left (i.e., will have a long left tail). A simple simulation confirms this assumption.

The bottom time series plot corresponds to the throughput produced from normally distributed K (number of users per cell) in the top time series plot by applying the $N_t \log \log k$ rule. The problem with this is the resulting distribution of throughput being skewed to the left (a long tail to the left, toward the smaller throughput values). This is not an expected throughput behavior, as we discussed in Sect. 2.3 (Fig. 6.16).

A similar transformation is effected when we apply the SSL.

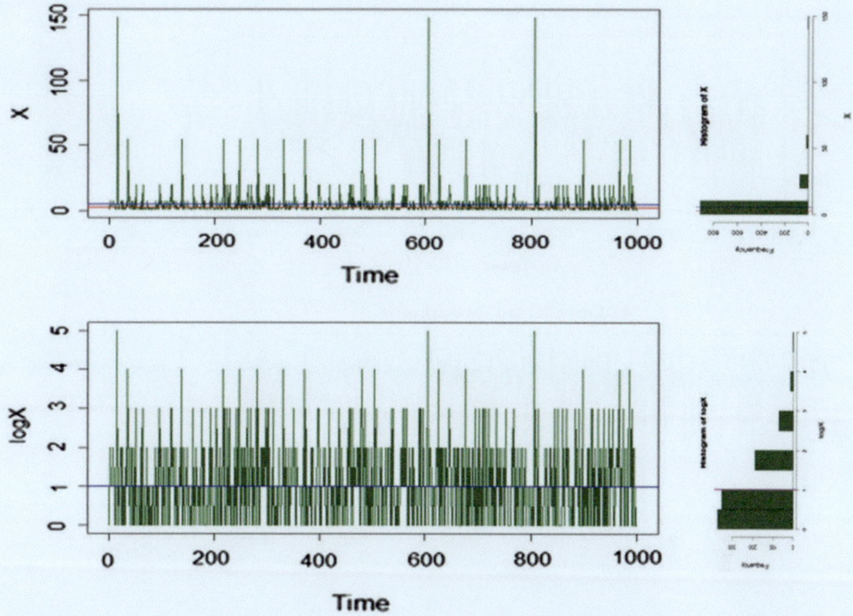

Fig. 6.15 Effect of a single logarithmic transformation on right-skewed data

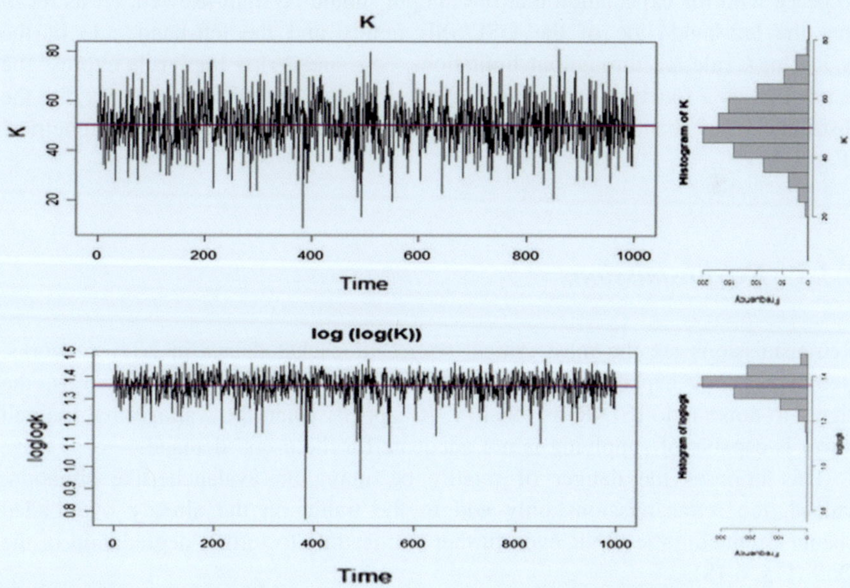

Fig. 6.16 Distribution transformation of the $N_t \log \log k$ rule (an illustration)

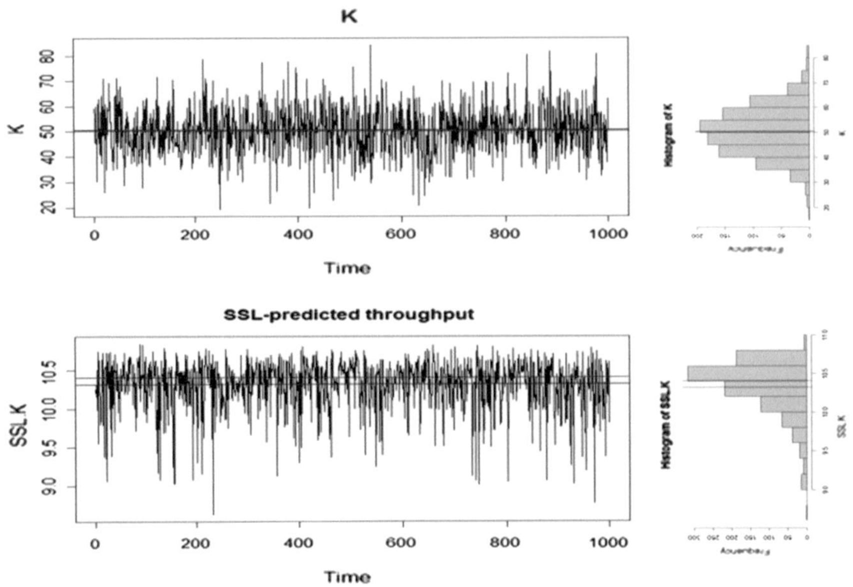

Fig. 6.17 Distribution transformation effected by applying the AISSL

Applying SSL also skews the distribution to the left. To bring these observations to peace with the expectation that throughput should be right-skewed, let us recall that the left-hand side of the USL/SSL family and the left-hand side of the $N_t \log \log k$ rule are throughput limitations: SSL and $N_t \log \log k$ rule express the bandwidth as a function of the number of servers/cell users. This ensures that the distribution of bandwidth utilization will be somewhat more symmetrical (Fig. 6.17).

6.3.3 Retransmissions

Retransmissions are the most critical source of packet delays in LTE networks. Retransmissions happen when the signal is too weak, or in other words, the signal-to-noise ratio (SINR) is too low. It happens when the wattage that the cell tower is capable of supplying is not adequate for the users' demand.

This imposes the danger of rapidly becoming an avalanche-like situation: Indeed, the retransmissions only add to the traffic on the already overloaded channel, reducing the SINR even further and leading to further degradation of the QoS (Fig. 6.18).

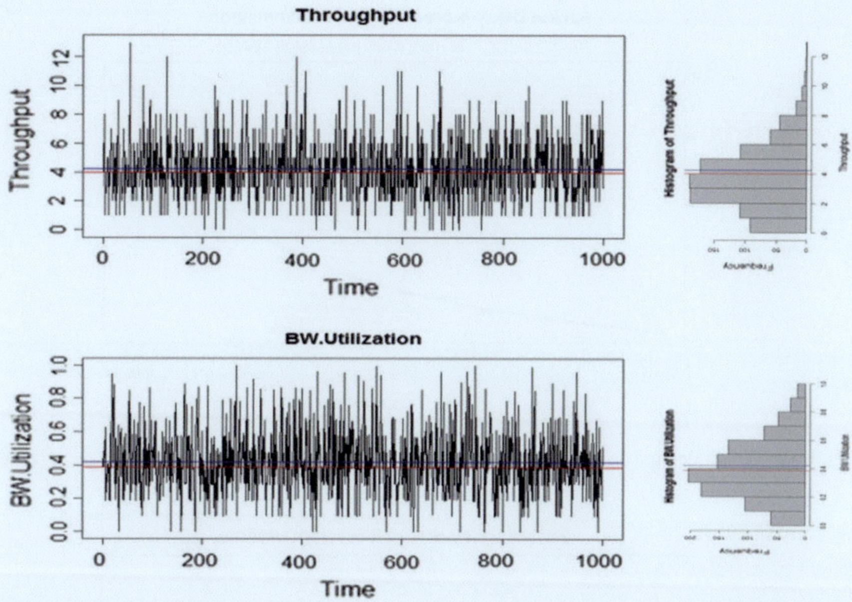

Fig. 6.18 Time series and distributions of throughput and bandwidth utilization. Note the significantly more symmetrical distribution of the utilization (the two charts on the *bottom*)

In addition, packet delays for retransmitted packets increase as well, as the latencies get added up for each packet that is getting retransmitted.

Mathematically, if

X_0 = packet transmission rate (demand);
P_r = packet retransmission ratio (probability of a packet getting retransmitted);
T_0 = expected value of packet delivery time when no retransmission is done;
r = number of retransmissions;
W = workload (concurrency) on the channel;

We can use Little's Law to express the number of retransmissions as follows:

$$r = X_0 * T(r) * P_r$$

Then, the delivery time for packets with retransmission is as follows:

$$T(r) = T_0 * (1 + r) = T_0 * [1 + X_0 * T(r) * P_r]$$

And, after elementary algebra,

$$\begin{aligned} &T(r)[1 - P_r * X_0] = T_0, or \\ &T(r) = \frac{T_0}{(1 - P_r * X_0 * T_0)} \end{aligned} \tag{6.3.2}$$

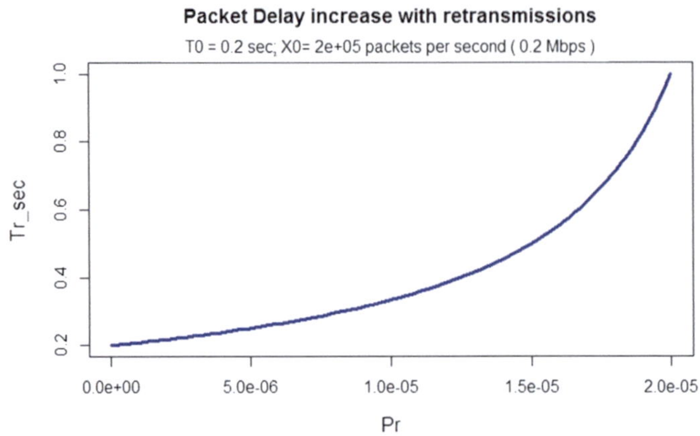

Fig. 6.19 Packet delay as a function of retransmission ratio

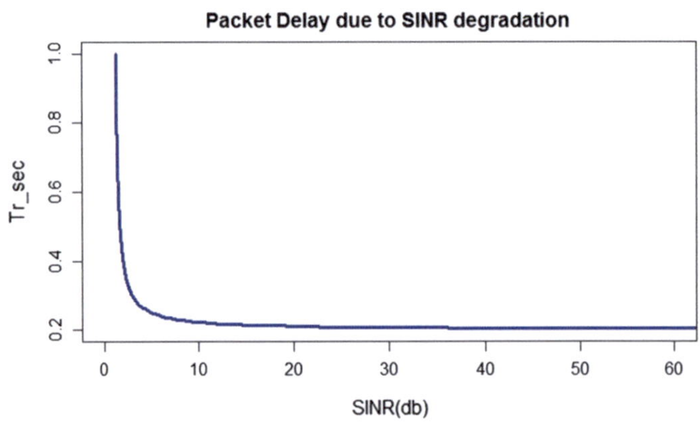

Fig. 6.20 Packet delay due to SINR degradation

As the retransmission ratio increases, the packet delay increases as well; at higher demands, the packet delay increases (Fig. 6.19).

In addition, P_r is a function of signal-to-noise ratio (SINR). SINR is a function of power provided and demand (number of users and their usage patterns), and if we plot packed delay as a function of SINR, accounting for retransmission, it will look as illustrated in Fig. 6.20.

6.3.4 *Router Load Balancing/Scheduling*

The two main load balancing methods implemented in routers are (i) round-robin (including weighted round-robin), where the current packet about to leave the router

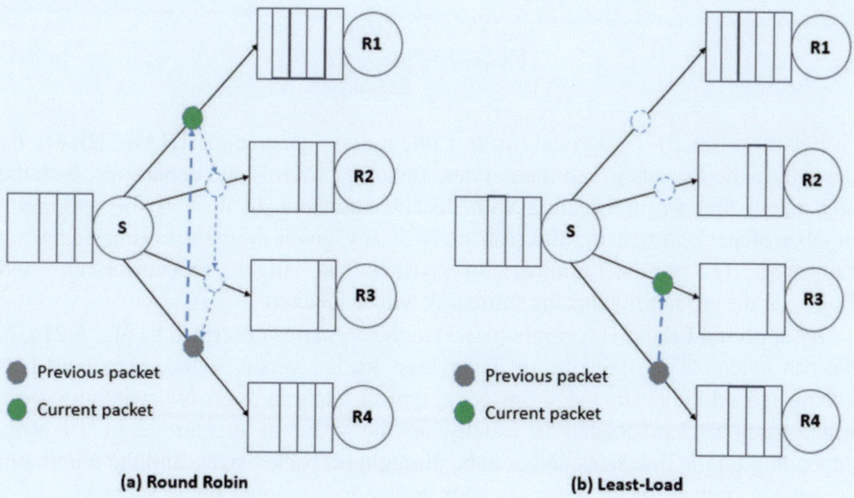

Fig. 6.21 Round-robin scheduling illustration

gets sent down the next outgoing link (aka degree) and (ii) load-based (least-Connections and Least-Loaded), where the current packet about to leave the router gets sent down whichever degree has the lowest load at the moment. Consider a router sending packets down two links, for simplicity (Fig. 6.21). Figure 6.21 illustrates conceptually round-robin (a) and least-loaded (b) scheduling. The black solid lines show the sending (S) and the receiving (R1, R2, R3, R4) routers and their corresponding links. The blue dashed lines correspond to the routing sequence.

With round-robin, regardless of how loaded the links are, if R4 has received the previous packet, the current one goes to the already overloaded R1, while the near-empty R2 and R3 are still near-empty. With least-loaded, regardless of where the previous packet landed, the next packet goes to R2.3. Least-Loaded routers are more complicated devices, as they need to maintain a picture of the world at least one hop downstream and always sort the links in the order of their current load, per the picture of the world that it has at the moment. It is usually done by pinging each of the downstream links and estimating the load on the link based on the RTT or the packet loss ratio. Queueing theory offers a way to compute the capacity required for each scenario in the form of Erlang-B and Erlang-C models (e.g., [FERR2012]).

If the total number of packets in flight is Q, then, assuming that arrivals are a Poisson (memoryless) process, and so is the process of passing packets through (processing), we can apply the Erlang-B model to compute the probability of the link being fully occupied at the time current packet arrives:

$$P_{\text{blocking}} = \frac{\frac{Q^N}{N!}}{\sum_{k=0}^{N} \frac{Q^k}{k!}} \tag{6.3.3}$$

Equation (6.3.3) is derived from Little's Law (see, e.g., [HARC2014]) by describing the stochastic and memoryless nature of arrivals and departures, with the adjustment that we are dealing with discrete numbers. In it, N is the maximum number of packets that the link can carry at any given moment in time (carrying capacity); Q is the number of packets in flight (concurrency); and P_{blocking} is the probability that the entire link will be blocked.

By applying Erlang-B formula (6.3.3) to the scenarios described in Fig. 6.21a, b, we can calculate the benefits of using least loaded versus round-robin load balancing (scheduling). In the industry, a typical capacity gain by switching from round-robin to least-loaded is usually on the order of magnitude of 10–30%, depending on the link sizes and counts, throughput, packet sizes, and the maximum acceptable probability of dropping packets due to all circuit being blocked.

One thing to watch out for when using load-based scheduling is the "black holing" of receiver routers. It happens when using least connections load balancing when a router that has just been put into operation in addition to the routers already taking traffic. Because it has not yet been in operation, its number of busy connections is zero, which means it will be targeted by all senders that have discovered it. In load-based operation, routers typically have a sweep rate parameter, which regulates how many rounds of pinging it can do every second. If the sweeping rate is not fast enough, the newly installed router can get overloaded and start dropping packets before the benefits of its being online start to manifest themselves. In the "compute" world, this usually happens with multicore load balancers, where cores are independent, and each core maintains its own view of the world. In the networking world, it can happen all the time when a new receiver node is discovered by more than one sender, forcing rerouting and retransmissions. The least loaded scheduling is somewhat more tolerant to black holing and other performance artifacts, as it takes into account the packet size, making the whole operation smoother.

6.4 Conclusion

We have demonstrated how to overcome the limitations of standard statistical tools and techniques in the real-world LTE network performance data and to extract useful information from such data for statistical process control, outlier detection and analysis, and queueing analysis. We also showed how advanced analytical methodologies can be used to advance our understanding of LTE network performance and lead to actionable insights. We intentionally left certain advanced techniques (e.g., quantile regression, principal component analysis, structural equation modeling, and Monte Carlo analysis) outside the scope, as the complexity of the mathematical machinery and practical implementations of such methods would have created a digression from the main topic of this book. Interested readers are referred to other literature.

References

[FERR2012] Ferrandiz J, Gilgur A (2012) Level of service based capacity sizing—presented at the 38th performance and capacity international conference by CMG (CMG'12). Published in the Conference's Proceedings

[SCHW2013] Schwarz S, Ikuno JC, Simko M, Taranetz M, Qi W, Rupp M (2013) Pushing the limits of LTE: a survey on research enhancing the standard. Access, IEEE, vol. 1, no pp. 51, 62 (2013). doi: 10.1109/ACCESS.2012.3.2260371, Url: http://140.98.202.196/stamp/stamp.jsp?tp=&arnumber=6514821&isnumber=6336544

[HETH2001] Heath R, Jr., Airy M, Paulraj A (2001) Multiuser diversity for MIMO wireless systems with linear receivers. In: Proceedings of 35th Asilomar conference signals, system computer, vol 2. pp 1194–1199

Chapter 7
Next Generation Network Performance and Fault Analytics Management Systems

7.1 Overview

The previous chapters described the various network segments of a typical LTE wireless service providers network, in particular the EUTRAN, EPC, EBH, Metro, and Core. These chapters provided the reader with a good background on the network itself, needed to develop meaningful and relevant models needed for network analytics. This chapter defines not only how the legacy systems differ from the proposed next-generation performance and fault analytics system (NGNM), but also practical aspects of operationalizing this NGPF analytics platform in a carrier network. This chapter will describe the aspects of a complete practical system suitable for an LTE carrier infrastructure and some practical approaches to realizing a system using an incremental approach. This chapter also describes the traffic collector that diverges from a legacy NMS system and its practical considerations for deployment in a representative LTE wireless service provider infrastructure.

Network performance evolved organically from performance management systems, and similarly network fault analytics evolved from network fault management systems. The early performance management systems simply reported and displayed the counters that were maintained by the network element, usually implemented on a separate control plane user space software that ran on the control plane processor. Fault management systems typically simply reported alarms emitted by network elements that know when a fault event was detected such as a down link or a vendor-specific threshold was exceed that triggered an event. An example of this vendor-specific event would be a vendor implementation of optical link monitoring and optimization. The vendor implementation may have proprietary software, and if the software crashes or uses excess memory, an event is triggered. This software is a vendor-specific threshold that triggers an event.

Performance analytic and fault analytics use complex automations and machine learning algorithms to deduce insights that typically take a significant amount of

D. Kakadia et al., *Network Performance and Fault Analytics for LTE Wireless Service Providers*, DOI 10.1007/978-81-322-3721-1_7

manual expertise and effort or time to go thru a vast amount of data. These insights typically provide valuable benefits to the organization such as identification by a senior network engineer to deduce the root cause of intermittent network issues, non-optimal routing, traffic engineering issues, bottlenecks, improved network transport load distributions, and other network optimizations.

Fault analytics can also include examples where complex root causes include transitory and intermittent symptoms and clues. Senior engineers manually correlate with other data sources that they have learned from experience to provide insights about the underlying causes. As an example that 1% of the network issues take 99% of the time and disproportionate operations costs. Automation using fault analytics leverage machine learning algorithms to handle not all, but a few recurring complex network issues, which otherwise would take significant human efforts to process and correlate vast amount of data to deduce root cause.

The first systems management tools were evolved along with computing systems such as IBM, which developed tools to trace network messages in order to pinpoint the sources of errors. As the number of computing and network manufacturers grew and standards-based protocols allowed interoperability, the need for multi vendor network diagnostic tools grew. Early network performance and fault tools consisted of either proprietary tools that were developed by the network equipment manufacturer who know their product best and which metrics to monitor, or the third-party vendors who focused on traffic when interoperability issues cropped up. Network tools for the enterprise are based on IP, whereas early telecoms network equipment were focused on the circuit-switched tools which were often developed by the network equipment provider themselves.

7.1.1 Network Measurements

Network analytics requires three fundamental categories of data measurements, which are normalized, time stamp aligned, correlated, and analyzed using statistical methods and machine learning algorithms to provide automated and unprecedented insights that are not obvious nor humanly possible due to the velocity and volume of data being processed. The three fundamental data categories include the following:

- **Network Traffic**—The traffic itself provides clues about what the subscribers actually experience from the time the packets leave the server and arrive at the user, or vice versa. These packets will experience changes that may be unexpected and require tuning the network and other non-obvious insights. There are many different traffic measurement approaches, each providing different perspectives and opportunities for analytics, as follows:
  - Offered network load—This can be a deceptively difficult measurement, because it represents the real offered load and the amount of traffic the source

wants to inject into the network. This is ideally what the network should be optimized in terms of raw bandwidth capacity. However in reality, applications are throttled, and the traffic is actually buffered at the source, so that the network has no way of measuring this, and often network capacity is based on what traffic was seen on the network after being throttled at the source, either in the TCP module in congestion control state or at the application blocked on a socket. Indirect measurements can be approximated using techniques such as interface utilization, netflow with adjustments to compensate for sampling, and other similar approaches.
- Packet latency on a per-QoS basis—This measurement can require tremendous amounts of processing capabilities and correlations as networks reach 100G and more when bonded or trunked. These measurements are key to optimize the QoS parameters of the network. Many network vendors are starting to expose KPIS, such as queue occupancy and drop rates, that start to expose some related measurements, but to measure the actual latency on a per-hop basis is difficult to obtain directly, and most feasible seems to be to infer this measurement by approximation and using indirect methods using network calculus methods.

- **Network Element State**—The network devices also provide clues about how it behaves and how it modifies the network traffic. Key measurements such as routing process resource consumption, CPU, memory, buffers, packet drops, interface utilization, and many others provide key measurements needed to indicate network element and its internal components' (hardware and software) state.
- **Network Transport and Topology**—Network traffic and network element state data such as netflow by itself do not provide a complete picture; this information must be correlated with the topology of the network to understand what the root cause of issues may be. Further, the network is constantly changing, so an automated process is required to stay in sync with changes in the network transport links, topology, and other transport-related configurations. There is a subtle distinction between the measurements related to transport and traffic. Traffic measurements are about the traffic in the perspective of the applications experience and are usually end to end. Transport measurements are about the transport elements and are usually point to point. The measurements in general usually provide insights about what can be inferred. If a network link is congested, it does not necessarily tell us anything about the application or user experience since their traffic may get rerouted, but it does tell the operator that the link needs to be increased in capacity. Similarly, if there is excessive packet delays, the transport links may actually be underutilized, but the router may have misconfigured the QoS parameters or packet scheduling algorithm, creating the latencies on a per-QoS class basis.

7.1.2 Standards

Network Management Systems (NMS) and Element Management systems (EMS) have grown significantly over the last decade in LTE wireless service provider network infrastructures due to the migration from circuit-switched technologies to IP and its associated management tools for performance and fault. The problem with this growth is that these tools have been historically designed for small- to medium-sized enterprise networks, and these systems do not scale to the size wireless service providers need. Further, service provider network managers purchase systems in silos, without any company wide structure or plan in particular without a multivendor-integrated NMS architecture. This model served the carriers adequately in the initial years when there were few systems and few vendors. However as the network scaled, in terms of network elements as well as EMS/NMS, this model resulted in inefficiencies, for example:

- Excessive EMS/NMS due to overlapping functionality;
- Redundant polling by multiple NMS systems on a particular NE; and
- Redundant copies of the same data in different systems.

PM and FM systems are now moving to functionally centrally manage all PM and FM aspects of the all the NEs in the network, ranging from the RAN, Backhaul, and Core network. This approach will reduce the number of redundant PM and FM functions in the carrier network and reduce the data integrity issues and NE load issues.

Further, increasingly carriers are requesting vendors to the following standards:

- EUTRAN—3GPP
- Optical—TMF
- IP—IETF

Standardized NGNMS PM and FM systems allow next-generation analytics that can inter-operate with different vendors' implementation on one component of the overall solution.

From an operational realistic perspective, next-generation standards-based PM and FM systems can be deployed in small phases, starting initially by pulling PM data from the existing EMS on their northbound and eventually the EMS can be bypassed altogether and pull data directly from the NE. Ideally, it would be desirable to rip out legacy systems and install new standards-based system that integrate with each other; however, the most complicated part of any carrier network change is the interim migration part, where you need to keep the network running at the same time as install new systems and carefully splice out the old systems. The next section will describe in further detail the practical aspects, including challenges and solutions to operationalize these new proposed platforms.

7.2 Network Performance and Fault Management Systems Architectures

A performance management system can be defined as the collection and production of performance counters of all the NEs in the carrier production network using industry standard protocols, message formats, and interfaces. Similarly, a fault management system can be defined as the receiver of network event messages triggered by network elements that detected a fault in the carrier production network using industry standard protocols, interfaces, and message formats.

In practical terms, operators realize the proposed next-generation performance and fault big data analytics (NGPM) architecture, which is not a simple upgrade but a complete fork lift replacement due to the completely different paradigms. Legacy systems are essentially based on some type of relational database, often a massive Oracle deployment which is fundamentally few but large servers, and the NGPM system is based on some variant of Hadoop deployment which is many small servers. As can be seen, the complexity lies in the migration path from the current NMS architecture to the next-generation performance and fault big data analytics architecture. This chapter will propose an incremental approach which will reduce the risk of deployment to the operator and incrementally splice in the NGPM system, and splice out the chunks of the legacy system component by component.

A typical legacy NMS architecture is shown in Fig. 7.1. As can be seen, network planning typically procures network elements from a vendor and then that vendor would also provide the NMS systems. As the carrier network grows with different vendors and different types of network equipment each with its own NMS system.

Figure 7.1 shows that the traditional NMS/EMS solutions were closely tied with the NE from the same vendor, which yielded closed proprietary systems. This model served use well in low-volume infrastructures. However as the number of NEs scaled and concurrently the data generated scaled due to tenfold increase in bandwidth, the NMS/EMS did not scale, resulting in expensive low-performing solutions. Often adding to inefficiencies, NEs were being polled for the same data by multiple different performance management (PM) and fault management (FM) network management (NM) systems.

One key challenge in the development of a production grade network analytics framework is that tremendous amounts of data must be processed and correlated. In practice, this is very difficult and most often not feasible with existing multivendor proprietary closed legacy NMS systems as shown above. There are several key reasons as follows:

1. The data models from one system to another are different, to remap data, often can only be accurate to 50% and that is with sizeable amount of manual efforts which usually requires consulting hours from the vendors that own the data, and these costs are often very high compared to the actual technical complexity.

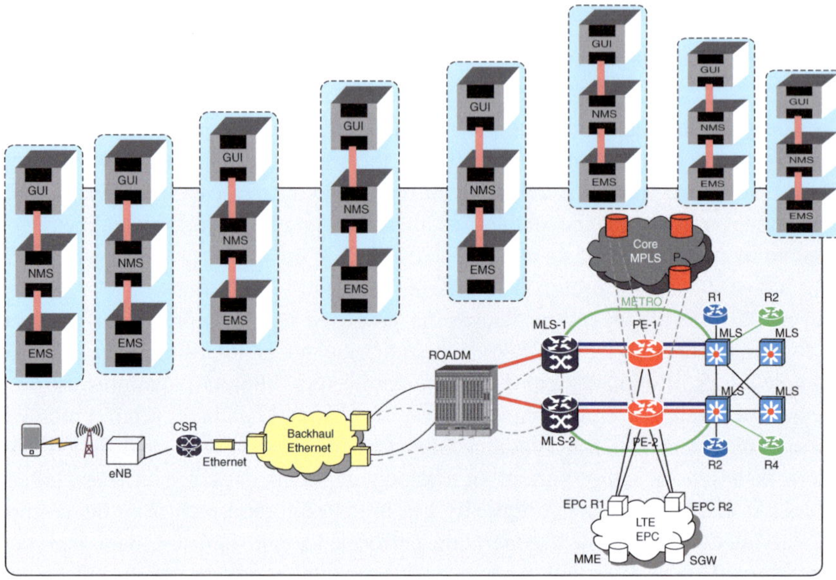

Fig. 7.1 Legacy carrier PM–FM architecture—silo-based—closed proprietary systems

2. The proprietary systems themselves do not scale. Near real-time data analytics platform will not work when the proprietary NMS system, going thru their APIs to pull data, cannot sustain a certain throughput.
3. There are multiple layers of inefficiencies such as the carrier has now multiple boxes to support and the data it self must go thru several layers before arriving directly to the network analytics platform.

The following section describes an approach to a practical transition, incremental and phased approach to migrate from a legacy to next-generation production grade network analytics platform.

7.2.1 Incremental Migration to Next-Generation Performance and Fault Analytics Platform—Integration of Existing NMS Systems at Top Layer

There are two equally valid approaches to incrementally move to a highly scalable network analytics framework based on big data analytics either starting from the top or the bottom. The choice will depend on the carriers' particular situation, goals, and constraints.

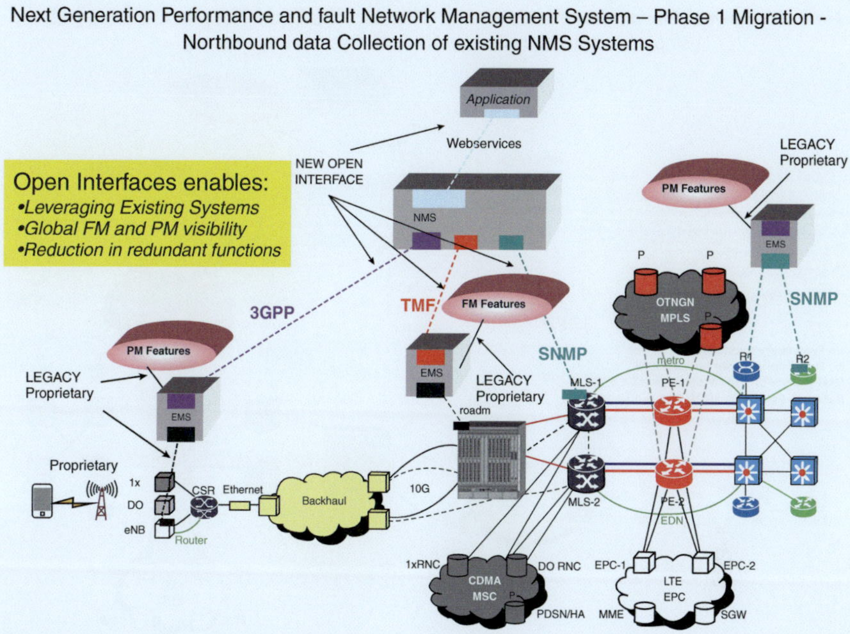

Fig. 7.2 LTE carrier top-down incremental approach toward next-generation network analytics architecture

We will explain the top-down approach; however, due to the modular characteristics of the architecture, it will be obvious that the same concepts apply to start from the bottom after reading this chapter.

Figure 7.2 shows an example deployment diagram of how the top-down approach can be executed. At the bottom southbound interfaces, there are proprietary interfaces that connect proprietary network management systems to the devices they are managing or collecting performance data. The northbound interfaces of these proprietary systems are almost always accessible to any third-party or in-house development software application. The top-down approach basically pulls data from all these closed systems from the top layer, and then the development of the network analytics applications can begin which interact directly with end users such as network operators, network engineers, network planners, and network monitors.

There are several benefits for this top-down approach. Firstly, without spending too much money, the functional viability is verified. If, for some reason, this whole project is found to not add sufficient value, based on end user feedback, it can be terminated immediately without massive disruptions to the lower layers. The lower layers require massive investment, because of the fact that the NE has been sized carefully, implemented proprietary data structures to be specifically polled by their respective proprietary systems, often in proprietary binary format. Opening the NE interfaces will take long lead times, discussions at standards bodies such as 3GPP performance working group for SAE.

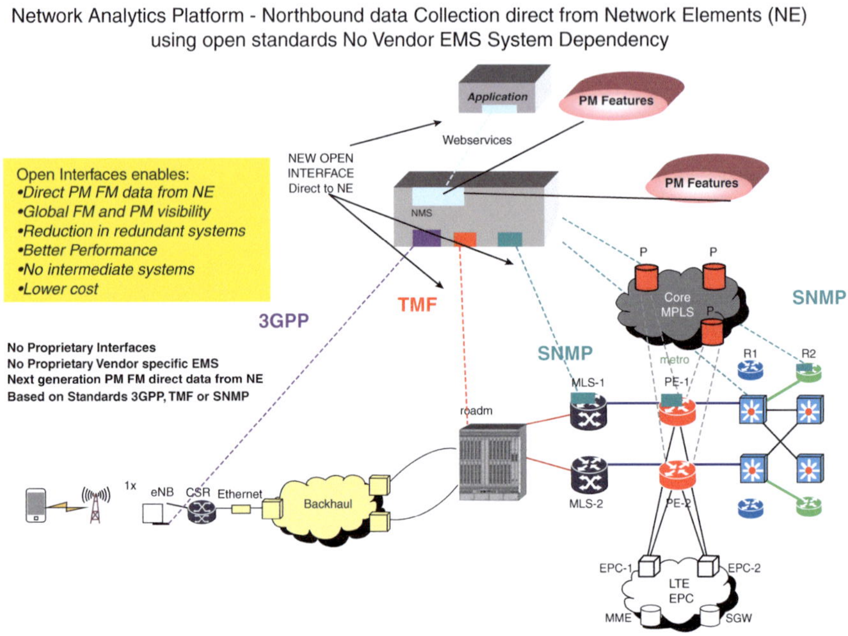

Fig. 7.3 Network analytics platform interfaces directly with NE using open standards

Figure 7.3 shows more details on the migration toward a long-term network analytics architecture. The overall end goal is for the network analytics platform system to pull data directly from the network elements using open standards—3GPP, TMF, or IETF. The vendor-specific network elements (NEs) have a vested interest in using proprietary protocols to send data to the higher cost, proprietary vendor-specific EMS, and then PM data forcing carriers to purchase the vendors proprietary system in order to access data for analytics. Due to the long lead time, Fig. 7.3 shows how the northbound interface of the vendors solution can be accessed using either a proprietary API or open standards interfaces. The idea is that if the NE vendors can open the northbound interface to open standards, then this same open interface can be argued to be implemented directly on the northbound interface of the network element itself, removing the dependency on the vendors' proprietary EMS.

Figure 7.3 shows an ideal situation where the network analytics platform accesses the NE directly using open standards; however, not shown is that in order to get to this point, it may be necessary due to timing that the access to NE direct may not be immediately possible, so an interim solution offered by NE vendors will most likely require access via the vendors' EMS to indirectly access performance and fault data from the NE.

7.3 Design Considerations for Carrier Grade Network Analytics System Optimized for Performance and Cost

As a consequence of the centralized architecture of the performance and fault network analytics system, the scale of the data increases exponentially and hence newer technologies need to be leveraged to reduce the cost and increase the performance. Big data architectures based on Hadoop combined with columnar databases is an ideal fit. Current commercial NMS systems are based on the relational databases which do not scale and are expensive. Moving to open source, Hadoop provides low cost and good availability. Hadoop is a file system and itself is not sufficient for a complete network analytics system, a database is still needed to organize and retrieve massive amounts of data. The nature of performance and fault data is structured, and it is time series. Columnar databases are a good fit since it exploits the fact that many variables are repeated in time slot measurement such as host name, interface name, IP address, and sometimes even the measurement itself.

The following section provides some insights into sizing and structuring the data for a network analytics (NA) system that is optimized for cost and performance.

The NA system architecture should be designed considering the types of workflows that will be forecasted in order to implement the network analytics applications. There are two fundamental types that impact the storage architecture.

1. **Ad hoc Queries**—The ad hoc queries perform predominantly random searches, and this work pattern is ideally implemented by relational databases, which should be indexed on fields that will be searched with the highest probabilities. Further, depending on the carrier architecture, a distributed database may be needed, which requires a costly join operation, which requires massive amounts of data to be moved from one database located in a certain region to another. Distributed databases are often needed in large carriers, since the majority of the data collection from the network elements must be local, so that massive polled data does not need to be backhauled to a central server. Depending on the analytics use cases, it may be necessary to weigh all the factors to develop an optimal architecture where some data to execute random queries may need to cached to a centralized server, and the low-probability ad hoc query data may be best left in the remote distributed location and fetch only as needed, keeping in mind the cost in terms of network bandwidth and latency to process that ah hoc query.
2. **Reports**—The reports are based on a priori-known queries and data sets. These types of queries are best implemented by columnar databases, where the fields of the known queries are sorted and retrievals are optimized. Additional benefits include that data can be compressed using techniques such as run length encoding. Columnar databases can still grow large; however, commercial products such as Vertical have proved to scale to petabyte sizes with good performance characteristics.

Generally speaking, most operators look at high granular data for troubleshooting purposes and only go back a maximum of 2 weeks. However, longer term network analytics need to consider data that span years. A common technique is to store granular data in fast expensive storage, such as SSD, and then roll up longer data for longer term analytics applications.

An example strategy is as follows:

- First, 2-week data—1-min granularity—raw data actual measurements—store in Fast SSD.
- 2 weeks to 6 months—1 hourly averaged and rolled up averaged data. This reduces storage by 60 times or one order of magnitude.
- 6 months to 2 years—daily averaged and rolled up averaged data. This reduces storage by a factor of 24 or one order of magnitude.

7.3.1 Carrier Grade Network Performance and Fault Analytics System Features and Functionalities

In this section, we describe features and functions that a typical LTE network service provider will require in a next-generation network performance and fault analytics system in order to realize a certain level of operational efficiencies.

- **Open System**—Description of open system data access and open standards-based application program interfaces and related protocols, and specifically not proprietary protocols. A detailed description of all protocols, message exchanges between manager and agent and remote network elements. This is required for interoperability with other systems to access data.
- **Remote Access**—Support may be required from the vendor into the carrier secure network to access the new Analytics system for routine maintenance and troubleshooting. The system must have a secure remote access support and must be approved by the Carrier Security Team.
- **Performance of System Itself**—Performance limitations of the analytics platform in terms of collecting, processing, and storage capabilities. All network analytics systems have limitations, and the key considerations are how many NEs can be polled and data received processed and stored per second. Poorly implemented systems in JAVA with an untuned open-source database will perform much poorer than a highly optimized C++ application with an in-memory commercial carrier grade database. There may be storage limits, for example, if the application is using 32-bit data structures and another application is using 64-bit data structures, there may be inherent size limitations.
- **Tools**—Supporting peripheral tools is also very important to the long-term growth of the network analytics system. Monitoring tools that clearly show the bottlenecks of the system, and diagnostic tools for trouble shooting as a

development support tools to compile MIBS and develop additional custom applications.
- **Standards**—Another related consideration is the support for standards-based interfaces and data models. It is of vital importance that the network analytics platform support an information model based on standards such as TMF and 3GPP.

When the information model is clearly specified and based on standards, it becomes much easier to integrate with other systems and operate and correlate data, which is one of the main foundations of network analytics.

The following are the 3GPP Release 9 PM specifications; at the time of writing, this was the latest commercially stable version.

- TS 32.401
- TS 32.404
- TS 32.425
- TS 32.426
- TS 32.450—E-UTRAN KPIs
- TS.32.455—EPC KPIs

7.3.2 TS 32.401—Performance Management Concept and Requirements

This section describes a condensed summary of TS 32.401—concepts and high-level requirements for the management of performance measurements and the collection of performance measurement result data across an operator network. It defines the administration of measurement schedules by the network element manager (NEM), the generation of measurement results in the network elements (NEs), and the transfer of these results to one or more operations systems, i.e., EM (s) and/or network manager(s) (NM(s)).

For the purposes of the present document, the following terms and definitions apply:

Element Manager (EM): Vendor element managers have in the past implemented all FCAPS functions. Now with the performance and fault being implemented by the operator, each vendor is only required to implement provisioning and configuration of the vendor-specific network element.
Network Manager (NM): NM provides a package of end user functions with the responsibility for the management of a network, mainly as supported by the EM(s), but it may also involve direct access to the network elements. All communication with the network is based on open and well-standardized interfaces supporting management of multivendor and multitechnology network elements.

The NGPF analytics system can be deployed at the network manager layer and will pull and or receive PM data directly from the NE. In order to reduce load on the NE, and in many cases other applications include the EM will require PM data, the PM and FM SYSTEM. will implement the 3GPP, TMF or IETF interfaces to provide the PM data on behalf of the NE. It should make no difference to the EM or any application that pulls PM data from the NE, whether it gets the data from the NE or the NGNMS PM system.

7.4 Network Traffic Analytics Architectures

In the above sections, we described NGPF solution architectures that were aligned with legacy NMS solutions architectures with the key common data source being the data from network elements. The proposed NFPF system has one significant additional data source—traffic. The collection of traffic is problematic due to scale, velocity, and volume. In Chap. 5, we describe the IP network architecture of LTE wireless service providers, and then we describe in particular traffic analytics—QoS and TCP. We are missing how all this data traffic is collected. We now address the collection of traffic.

In this section, with the intent to complete the NGPF solution architecture, we now describe how traffic collection will fit into this new model.

Figure 7.4 shows one proposed solution example where a nationwide carrier deployment of a traffic analytics solution consists of regional or local collectors that feed summarized results into a global traffic analytics system that stitch together many regional views into a holistic cohesive nationwide view of network traffic

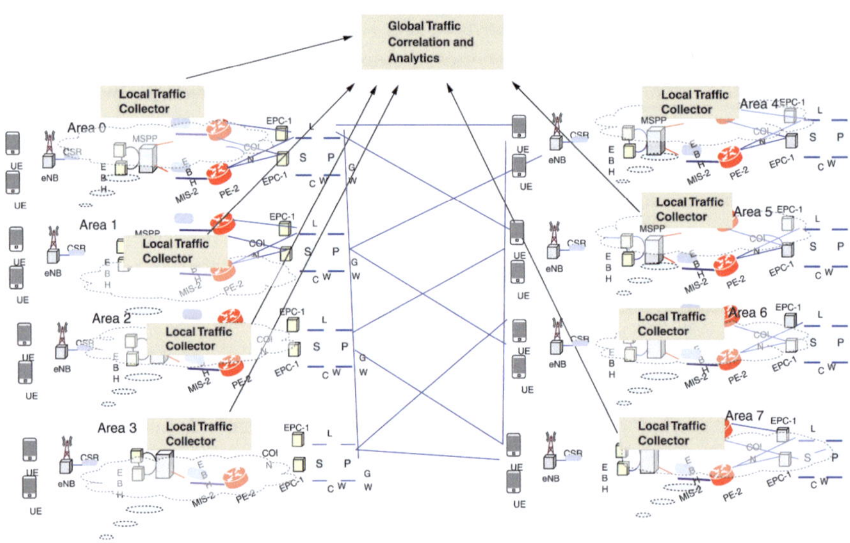

Fig. 7.4 Scalable hierarchical traffic analytics system for LTE carrier

insights. From this framework, a dashboard can easily be produced, which provides a high-level summary of all key segments that correlate with the segments as shown in Fig. 7.4. A typical LTE wireless provider will have similar network and components in various parts of the country geography. A dashboard summarizes on various timescales the best, average, and worst latencies on a per-segment, per-DSCP, or queue basis, based on the measured customer traffic.

7.5 Network Analytics Engine

In the previous sections, we described potential NGPF solutions in terms of data collection, storage, and migration from a practical and high-level perspective. In this section, we describe in a little more depth the analytics engine.

This section describes an example network analytics system that is optimized for LTE network analytics. Current network management systems for performance monitoring are designed using legacy general purpose relational database management systems. These legacy systems do not perform well due to the fact that LTE nodes, such as eNodeB, produce tremendous amounts of data, and the data are sent in XML files. The relational database architecture is optimized for writes, but not reads. Performance and fault analytics systems work loads are write once and read many times. In order to design an analytics system that will scale and perform within reasonable time frames, a suitable system will need to ingest massive amounts of data, yet process massive volumes of data and correlate different sources of data.

This section describes an approach that is optimized for LTE usage patterns, which is write once and read many times, but also covers two distinct work patterns:

1. **Real-Time Mode**—A separate subsystem within the proposed architecture that is optimized for real-time performance monitoring, critical for tactical trouble shooting, and operations urgent problem resolution issues. Characteristics include shorter time frames, fast response, and less memory requirements.
2. **Batch Mode**—A separate subsystem within the proposed architecture that is optimized for massive reports and analytics processing, used for generating reports, crunching massive amounts of data, and processing for planning within the enterprise. Characteristics include longer time frames, slower response, and large memory requirements.

Due to the two diverse use cases and accordingly vastly different system requirements to meet these two use cases, two separate subsystems will be proposed that are optimized independently, with the end result being a system that is overall optimized for both vastly different requirements.

7.5.1 Current Approach

Current approaches typically include a relational database management system to store performance data in tables. These systems typically store records and its related attributes in a sequential manner to a disk, in order to minimize disk write head movement. In order to read, this approach suffers from disk read head latency (seek and rotational latency). This approach was designed for enterprise systems, where the data is not read often but written often.

Network management systems, in particular performance management systems, are typically time-series polled data, written only once, never updated, other than deletion where data is often summarized or rolled up into summaries. Data is often read many times by different users for different purposes. LTE introduces new challenges, which are massive amounts of data. Figure 7.5 shows a simplified typical centralized approach for a performance management system using a RDBMS approach.

7.5.2 Limitations of Current Approach

Typical RDMS start to encounter major latency issues as the number of tables and sizes scale to accommodate the amount of performance data that is generated from LTE components. RDMS must often perform joins, which involve computing a Cartesian product, which requires fetching data, combining, and then filtering.

1. Latencies for random reads;
2. Centralized bottlenecks for processing and computing analytics; and
3. RDMS systems typically scale vertically which involve expensive centralized servers such as Sun servers.

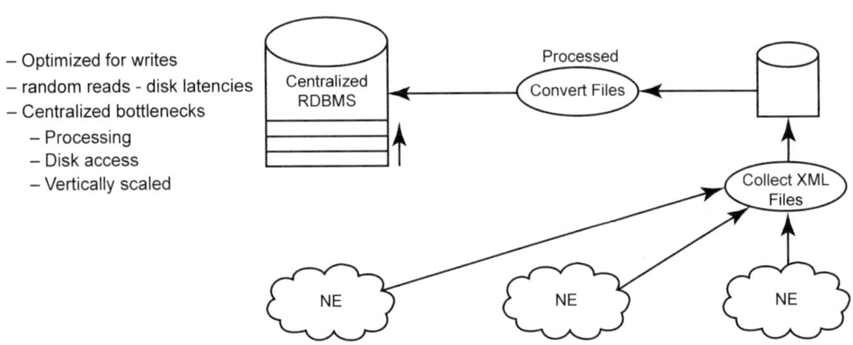

Fig. 7.5 Legacy performance management system based on centralized relational database management system

7.5.3 *Proposed Approach*

Figure 7.6 describes the proposed approach based on two subsystems: one optimized for short-term and real-time performance data feeds, and one optimized for long-term performance feeds in batch mode.

On the left side, we have the real-time subsystem. The overall architecture leverages the Hadoop data and name node architecture, where the name node interfaces with the client and manages the data node that actually store the data.

On the right side, we have the batch subsystem.

The client will interface with the real-time or batch API, and the appropriate subsystem will be invoked to provide the appropriate data. For example, the RT

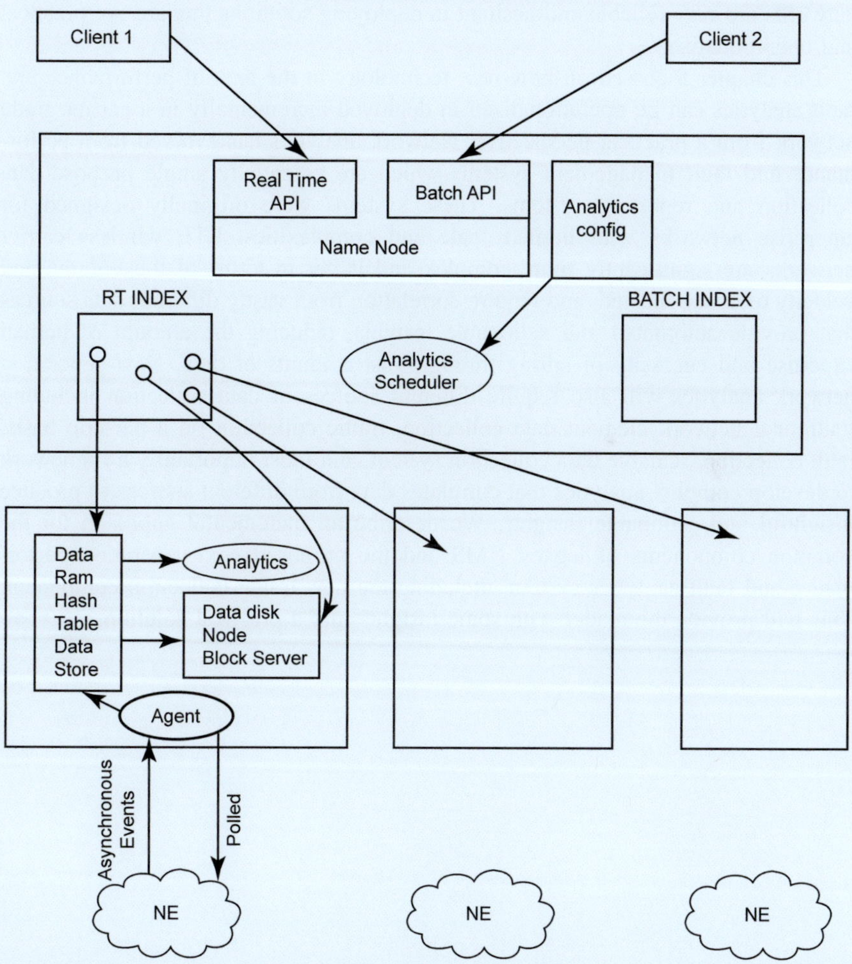

Fig. 7.6 Proposed scalable and flexible network analytics platform

index will quickly identify which data node actually stores the data and then go to the data node hash table data store to further find the actual data stored at the data node.

Figure 7.6 shows how the data is physically stored originally and the amount of compression gain by exploiting the fact that there are predominate number of zeros.

7.6 Summary

LTE wireless service providers like most operators are very conservative and are interested in real solutions. Performance and fault analytics theory into practice, particularly due to carrier scale and velocity, is always a major issue. Carrier networks are very reliable and hesitant in deploying solutions that are not practical and operationalized.

This chapter focussed on how new technology in the area of performance and fault analytics can be operationalized in deployed incrementally in a carrier grade network from a practical perspective. Network analytics has evolved from performance and fault management systems which are essentially single purpose data collection and reporting systems. These systems were originally designed for enterprise networks with limited scale and complexities. LTE wireless carrier networks are significantly more complex and larger in terms of the volume and velocity of data collected, and require correlation from vastly different data sources that provide automated and actionable insights, reducing the amount of human expertise and necessity of sifting through vast amounts of data. Next-generation network analytics will also require different sources of data collection including traditional network element data collection, traffic collection on a per-hop basis, fault collection, scalable data collection system, and most importantly a framework to develop complex analytics that correlates data from different sources to produce insightful and actionable insights. We describe an incremental approach for the common components of legacy NMS and the proposed next-generation performance and fault analytics system, including the new traffic collection component. This will provide the reader with some insights into a practical solution.

Chapter 8
Self Organizing Networks (SON)

Self-organizing network (SON) enables wireless network operation efficiency and automation. Network performance and fault analytics are the foundation for self-organizing network (SON). As wireless network traffics exponentially grow, network operators are accelerating network densification, introducing varieties of radios in the system. The cellular network is moving from simple homogeneous macro only network to a more complicated heterogeneous network. This includes small cell, micro, pico, remote radio head, distributed antenna, enterprise, and residential femtos. Services are expanded from traditional voice and data only to voice, video, machine, and others. Network operators are also deploying more spectrums, from licensed cellular 850, PCS, 700 MHz, AWS, to unlicensed spectrum on 2.5/5.8 and 3.5 GHz shared spectrum.

Self-organizing network (SON) is essential to build and operate this type of denser network with various types of radio nodes and spectrums. It boosts capacity and performance, reduces interference, and improves mobility. It touched every aspect of wireless operator operational organizations, including RF design, service performance, engineering, and operation. SON becomes an efficient engine to glue organizations together.

SON started in 2007 with concept defined in 3GPP. LTE is the first access technology with SON standardized in air-interface protocol. Industry has started use of the SON concept since introduction of LTE commercial service in 2010. SON has evolved into a multiple vendors, multiple technologies, and multiple layers solution. It significantly enhanced traditional PM/CM/cell traces to a full network optimization and visibility. With soft-defined network (SFN) and network function virtualization (NVF), we are expecting more dynamic end-to-end service delivery intelligence and flexibility.

In this chapter, we will introduce SON in Sect. 8.1 and SON architecture in Sect. 8.2. We will present major SON optimization features in Sect. 8.3 and provide some practical use cases and deployment guideline in Sect. 8.4. Chapter summary is given in Sect. 8.5.

D. Kakadia et al., *Network Performance and Fault Analytics for LTE Wireless Service Providers*, DOI 10.1007/978-81-322-3721-1_8

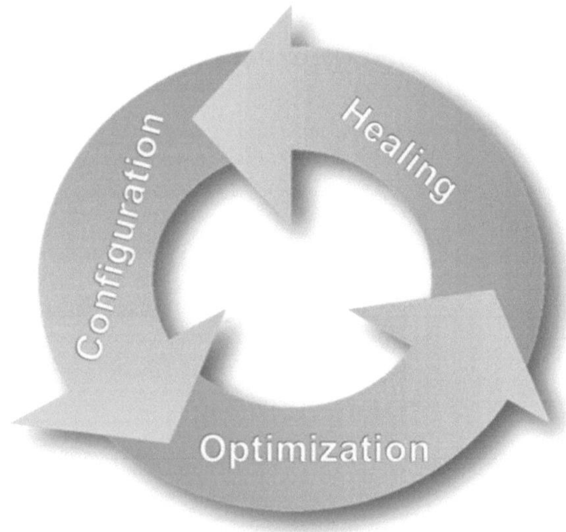

Fig. 8.1 Basic components of self-organizing network

8.1 SON Introduction

Self-organizing network includes 3 aspects of network engineering: self-configuration, self-optimization, and self-healing. Self-configuration means Plug-and-Play. An eNB can automatically turn on without technician assistance. Self-optimization means near real-time optimization and avoids manual intervention. Self-healing means the system can detect and recover from outage automatically (Fig. 8.1).

Those 3 aspects are intervening with each other to ensure an efficiency operation.

8.1.1 Self-configuration

Wide-area network has more sophistic processing to turn on an eNB than a home Wi-fi router. Traditionally, a technician will bring a computer and a script to run at the site to configure an eNB and bring a site alive. This could take hours or days. It includes firmware and SW version updates, cell site identifications, and IP address configuration.

With the self-configuration functionality defined in TS32.501, this processing can be automated. For example, the following procedure can be carried out to establish an eNB communication:

(1) eNB coming from the factory with an HW serial number/bar code
(2) When eNB is installed, the installer scanned and sent bar code and location of the base station to network management center

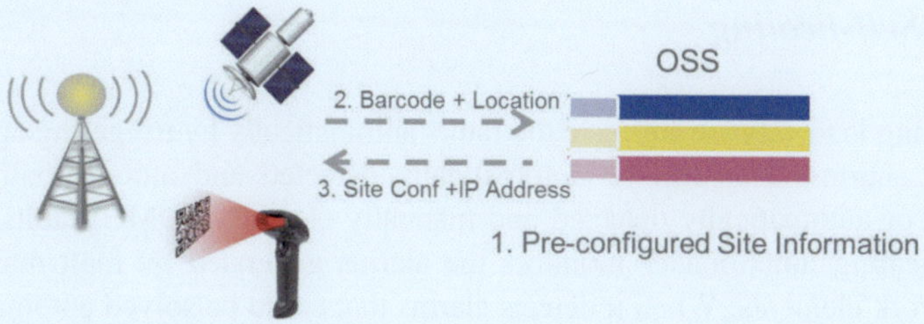

Fig. 8.2 Self-configuration procedure

(3) The OSS will link the bar code with prepared configuration
(4) When the eNB is power on, the site configuration and cell parameters will be sent to eNB. This includes the latest software and firmware updates
(5) The eNB is in operation (Fig. 8.2)

The IP address can be allocated with DHCP server associated with OSS server to ensure reliability. This entire processing only took about 15 min. With site installation file, site basic file, site equipment file built ahead in base station database, as well as defined at eNB, OSS, S1, X2, transport network.

Self-configuration reduces potential human error during site installation. Working together with self-optimization and self-healing, the self-configuration also can improve network performance the moment the site is turned on.

8.1.2 Self-optimization

LTE network intelligence comes from fine-tuning of various operational parameters. With the introduction of voice, video, machine communications and other service-specific features, network optimization becomes more and more sophistical.

Self-optimization will reduce the burden of engineering; in particular, with optimization involve a cluster of cells with multiple relationship among cells.

On top of operation efficiency, SON has demonstrated around performance improvement on 4G LTE network by adjusting parameter on a cluster level on top of cell site level. In a typical network, access failure rate and drop call rates can be reduced by 10%, while user throughput can be increased by 10%. This improvement could be even higher in a poor engineered network, while lower in a well-optimized network.

Major self-optimization functions involve automate neighbor relations (ANR), mobility robustness optimization (MRO), mobile load balancing (MLB), coverage and capacity optimization (CCO). The ANR is defined in TS32.511, while other self-optimization functions defined in TS32.521. ANR is used as an example for hybrid SON in Sect. 8.2, while MRO, MLB, and CCO are going to be discussed in details in Sect. 8.3.

8.1.3 *Self-healing*

Self-healing is to solve or mitigate the faults automatically by triggering appropriate recovery actions. This can be automatically detected and automatically cleared (ADAC) or automatically detected and manually cleared (ADMC) faults.

Self-healing functionality monitors the alarms generated by fault management and network elements. When it detects alarms that could be solved automatically, it gathers more necessary information (e.g., measurements and testing result) and does deep analysis, and then according to the analysis result, if necessary, it triggers appropriate recovery actions to solve the fault automatically. The trigger from fault management could be in minutes and took longer time to recovery. However, the triggers from network elements, such as eNB, require more rapid response to heal the fault automatically. 3GPP defined self-healing requirement in TS32.541.

Software faults can be recovered through system initialization at different levels, reload of backup software, activation of a fallback software load, download of a software unit, or simply reconfiguration. Hardware fault recovery depends on the existence and type of redundant resource. If the faulty resource has no redundancy, the recovery actions may be as follows:

(a) Isolate and remove the faulty resource from service so that it does not disturb other working resources;
(b) Remove the physical and functional resources (if any) from the service, which are dependent on the faulty one. This prevents the propagation of the fault effects to other fault-free resources;
(c) State management-related activities for the faulty resource and other affected/dependent resources;
(d) Reset the faulty resource;
(e) Other reconfiguration actions, etc.

If the faulty resource has redundancy, the recovery action shall be changeover, which includes the action (a), (c), and (d) above and a specific recovery sequence.

The self-healing functionality also monitors the execution of the recovery action/s and decides the next step accordingly. After a self-healing procedure ended, the self-healing functionality shall generate and forward appropriate notifications to inform the IRP manager about the self-healing result and all the information of the performed recovery actions may be logged.

8.2 SON Architecture

SON features can be classified as distributed SON features resided at eNB, and centralized SON features resided at NMS, at or above OSS level, as shown in Fig. 8.3.

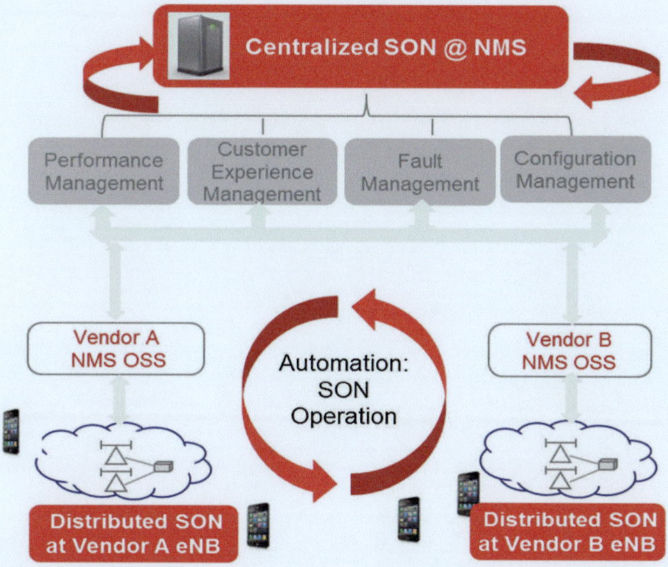

Fig. 8.3 SON architecture

Distributed SON will make fast decision based on eNB feedback in ms resolution. It is activated in a very fast past and more localized, for example, UE-based ANR feature. While centralized SON will make decision mostly based on information from configuration management, performance management, and cell trace. They are typically in 1 s or larger interval. For example, distance-based cluster level ANR filters.

LTE eNB can support both centralized and distributed SON functionalities. Therefore, an optimized SON solution shall be a hybrid solution, with both centralized and distributed functionalities. 3G HSPA only has centralized SON functionality. Hybrid SON communicated with performance, configuration, fault, and customer experience managements. And coordinates among sectors in a slow pace to optimize otherwise static parameters, such as antenna tilting, and load balancing across cells. It also supports multi-vendor interfaces.

8.2.1 eNB SON Functionalities

eNB SON functionalities require fast feedback. They are depending on real-time feedback from devices, such as, RRC connection messages between eNB and UE, radio conditions from UE. The feedback could be handover messages in 10–100-ms interval, or channel quality indication (CQI) at TTI level, e.g., 1 ms for LTE.

ANR is an eNB-based feature that is defined in 3GPP. UE will report strongest PCIs above a threshold. If the PCI is not in the neighbor relation, eNB will order

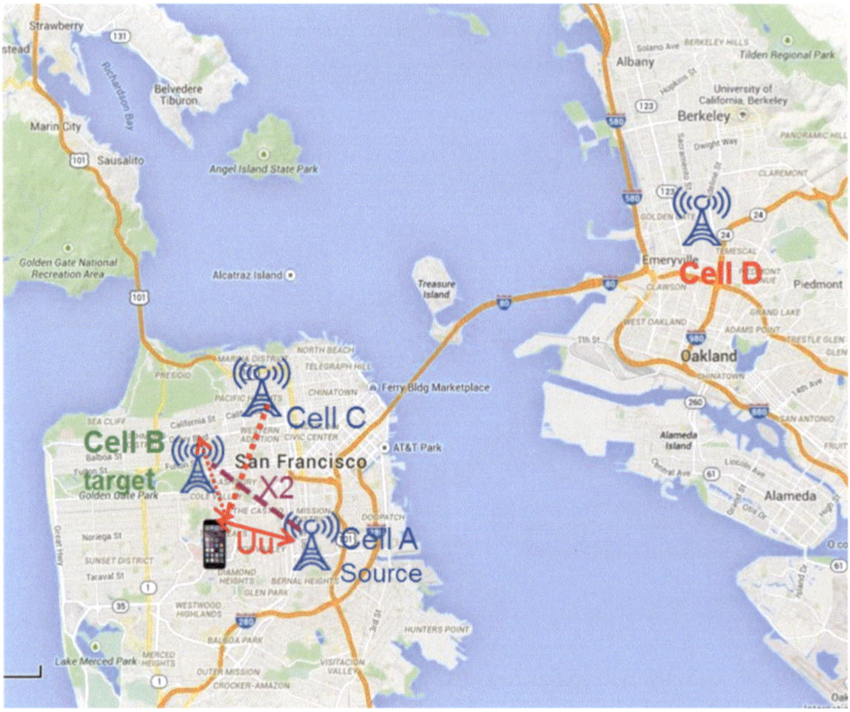

Fig. 8.4 Hybrid SON: ANR example

UE to measure Cell Global ID and eNB establish X2 with the sites, and OSS update the NR list. The eNB can immediately grant UE to handover to that missing neighbor cell with no call drop. This entire processing takes about less than 100 ms; it can only be executed at distributed eNB level. It is a function requires interworking among UE and eNB to ensure a smooth operation to avoid call drop to an unknown cell (Fig. 8.4).

8.2.2 Centralized SON Functionalities

Centralized SON functionalities are functions that would require cluster level information. Those would need coordination and relationship among cell sites in a much more large time interval. Current performance management typically pulls data in 15-min interval, while call trace in 1-min interval.

Centralized functions will take into account cluster information to optimize the parameter from an overall network perceptive. There are new functions, such as load balancing among cells and antenna tilting optimization. They can also further improve eNB functions, such as centralized ANR and MRO functionalities to further improve eNB ANR and MRO, respectively.

In an ANR operation, centralized function can further improve the ANR feature at eNB. They could oversee the eNB ANR functions, automatically remove non-used neighbors or add must have neighbors. One use case is a distance neighbor overshadowing nearby neighbor and is wrongfully detected by UE. This could end up with ping-pong between source and target cell sites or cause call drops. The centralized function can remove or deny this neighbor from the neighbor relationship, update blacklist to prevent this happen. The geo-location consideration here also can detect PCI confusion and collision.

Centralized SON can define parameter boundary and take advantage of cell site relationship.

8.2.3 Hybrid SON Control Mechanism

Hybrid SON combined both eNB and centralized SON functionalities. It is essential to ensure overall optimized performance under operator's policy control. Most current product implementations require disabling one set of features while the other is activated.

Both eNB SON and centralized SON are need to ensure proper performance of network. For example, with ANR at eNB, we cannot automatically add new sites and avoid call drops. Without centralized SON, we may experience more ping-pong and also lead to drop calls.

SON can operate on closed mode, where automation and parameter changes being enabled automated and open mode, where system performance engineer enables the parameter changes as need. For most handover optimization, closed mode could be more suitable. But e-tilting may be more suitable in open mode operation as the frequency of the changes could reduce the accuracy of the tilting control, in particular involve angels or position changes. This could eventually migrate to more frequent changes when more reliable pure electrical tilting mechanism in place.

To ensure proper network call connectivity and handover performance, both eNB and centralized SON have to coordinate closely together.

8.3 Self-optimization Features

Self-optimization is the center piece of self-organization. The objective of self-optimization is to maximize network capacity without performance and quality degradation, minimize congestion without losing coverage, filling coverage holes without overshooting.

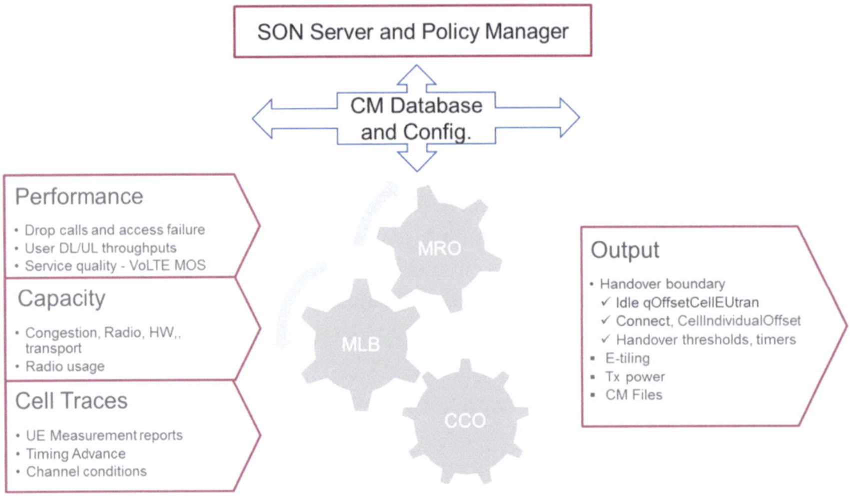

Fig. 8.5 Self-optimization inputs and output

eNB has built-in optimization feature using intelligent scheduler to optimize layer MCS and RB allocation based on channel quality indication. Here, we emphasize on hybrid SON to further optimize overall network performance and efficiency.

The hybrid SON features are adapted every 1–15 min based on PM data or cell traces. It is adaptive to change traffic patterns within the day. It can make adjustment to out of service eNB and react to both scheduled, such as Super Bowl, and no scheduled, such as traffic accidence. Hybrid SON also improves overall network efficiency and under-utilized eNBs.

The major inputs to self-optimization are KPIs critical to operators, and outputs are adjustment needed to the network (Fig. 8.5).

Major hybrid SON features are load balancing, coverage, and capacity optimization.

8.3.1 Mobility Robustness Optimization

Mobility robustness optimization (MRO) is used to resolve handover too early, too later, or handover to a wrong cell and too much ping-pong among source and target cells.

If the handover parameter setting is too conservative, the handover could be triggered when the signal strength of the serving cell is already too low or may not be triggered at all and caused a radio link failure. This is a common scenario in areas where user mobility is very high, such as along highways or train lines.

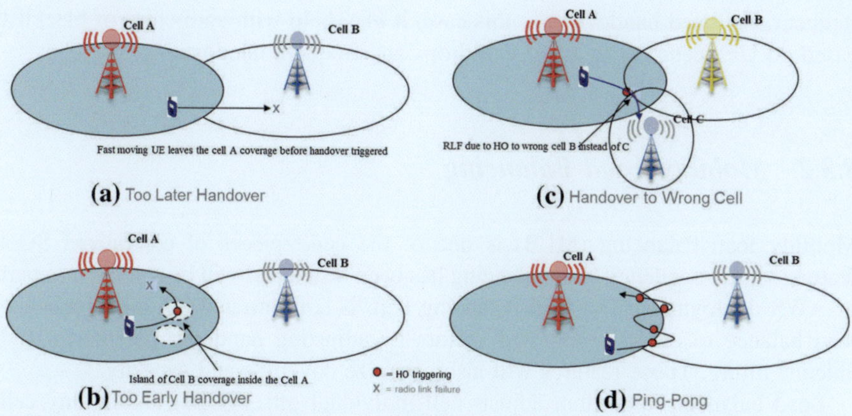

Fig. 8.6 Mobility robustness optimization (MRO)

If the handover parameters are too aggressive, the handover could be triggered when the unintended island of coverage of the targeted cell inside the intended coverage of serving cell. When the UE exits the island of coverage of the target cell, it cannot acquire the target cell anymore and lead to potential radio link failure (Fig. 8.6).

Not optimized handover parameters could also cause UE to handover to wrong cell. For example, UE is moving toward C, while handover to B. Then, a radio link failure happens, and UE reestablishes connection with C.

Operator can also set proper handover optimization objective to avoid excessive handover ping-pongs. Handover is a resource-consuming and therefore costly to the network operator. Figure 8.8d illustrated this ping-pong handover scenario while due to a combination of user mobility patterns and cell coverage boundary layout, UE experienced frequent unnecessary HOs that consume NW resources inefficiently. These scenarios can be remedied by handover parameter optimization, but this procedure is complicated as the goal of reducing unnecessary HOs can sometimes be conflicted to the goal of reducing the number of HO failures.

On top of handover parameter setting in connected mode, incorrect cell reselection parameters setting in idle mode may result unwanted handover right after RRC connection setup; HO parameter optimization function should also aim at detecting misalignment between cell reselection parameters and handover parameters setting and adjust the parameters to avoid such scenarios.

Centralized SON can utilize geographical knowledge and longer term KPI Observation to further optimize handover parameters. For example, it can improve handover performance of fast moving UEs on high mobility cell. In the highway cell, the handover could last too long for UE to handover and cause the drop call. cSON can introduce a negative CIO and ensures handover earlier in a highway cell based on long-term traffic pattern and UE current speed. In an urban environment, mobility setting can also impact highway site and static users in building, cause frequent oscillated handovers. In this case, it may help with auto-entry of blacklist based on UE scenarios to avoid call drops and frequent handovers.

8.3.2 Mobility Load Balancing

Mobility load balancing (MLB) is one of the centerpieces of centralized SON features. Inter-frequency load balancing has been tested and will be activated as part of AWS deployment. The load balancing here is concentrated on intra-frequency load balance to offload congested sectors by adjusting handover parameters and antenna tilting. Those features will also optimize coverage and capacity.

Load balancing algorithm adjusts cell individual offset (CIO) among any cell pair to adjust cell boundary when one site is congested. The congestion can be measured by number of connected users, downlink and uplink PRB utilization, and DL and UL user throughputs. When a congested cell is detected, the algorithm will analyze surrounding neighboring cells for potential offload. It then changes cell individual offset and qOffset to move the cell reselection and handover boundaries to shift traffics to neighboring cells. The offloading will dynamically vary and adapt to ongoing KPI feedback and traffic changes.

Figure 8.7 gives an example of load balancing. In this example, the red site is congested with 5% congestion rate. While surrounding site, e.g., yellow is not congested, and blue is congested at 1%. The algorithm is going to adjust cell individual offset to let the UE state in yellow cell longer, in this cause, shift the UE

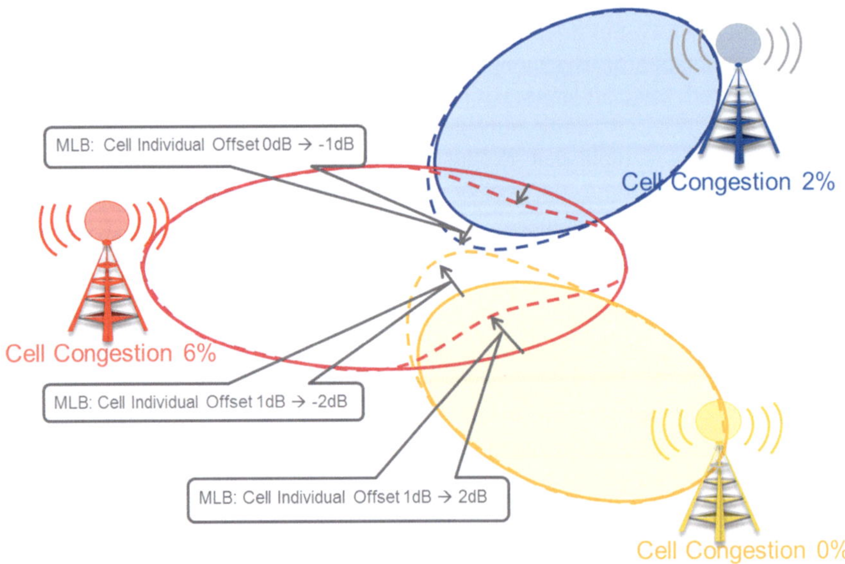

Fig. 8.7 Illustration of mobility load balancing

to yellow site 2 dB early or stay in yellow site longer. That is instead of moving UE to red site when its RSRP is higher than yellow site, wait until red site RSRP is 2 dB higher than yellow site.

These can very efficiently moving the cell boundary form solid to dash line and reshaping the UE distribution in this cluster taking advantage of geographic knowledge about overlapping cells and longer term KPI trend observations to detect traffic patterns. However, this centralized load balance feature cannot act alone. The same CIOs also are dynamically optimized by eNB for mobility robustness optimization (MRO). The final parameter adjustment needs to take into account both eNB and centralized SON adjustment simultaneously. At eNB level, inter-frequency and inter-technologies handover thresholds will also be enabled to move the traffics evenly across multiple frequency bands and multiple technologies.

Therefore, a successful load balance will take advantage of inter-frequency, inter-technology load balancing at eNB, and intra-frequency load balancing across neighboring cells within the frequency bands. It also needs to work together with MRO and ANR to ensure mobility robustness and proper neighbor relationship adjustment.

8.3.3 Coverage and Capacity Optimization

Coverage and capacity optimization (CCO) is to optimize cell site transmit power, antenna tilting, and others to relieve congestions and fill coverage holes. Those changes could also impact ANR, MRO, and MLB. Those features need to act coordinately.

LTE has frequency reuse of one and only support hard handover with no macro-diversity among neighboring sectors. Therefore, bordering interference control is essential to ensure cell-edge performance. Coordinated multiple-points transmission (CoMP) and inter-cell interference coordination (ICIC) are arming to improve the cell-edge performance. The best way is to control the interference by optimizing transmit power, antenna directions and tilting so that it only radiated to the intended coverage area, and restrict interference.

When the cell is overshooting, it not only creates interference to surround cells on downlink causing reduced SINR and quality of services, but also increases device transmit power, led to increased UL interference and power consumptions.

Detection and determine coverage hole or overshooting dominated in a troubled area is the first step toward coverage and capacity optimization. It is important mechanism adhering to operator predefined policy and parameter boundary. Those thresholds and boundary are defined and determined by an operator's operation philosopher and priority.

The over-shooter is determined by how far the radiated transmit radio reached. A CCO typically evaluates each cell to determine its 1st, 2nd, 3+rd tiered cells based on UE timing advance report. The over-shooter could be detected when the cell reached 2nd or 3+rd tier of cell coverage areas that caused KPI degradation.

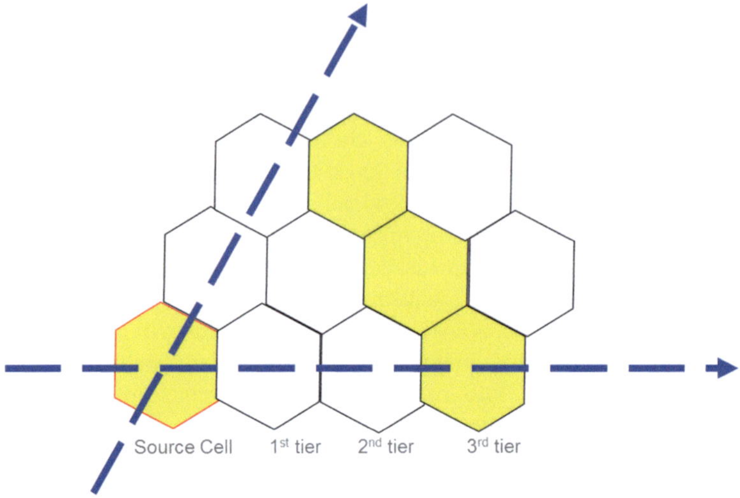

Fig. 8.8 Remote electric tilting

High number of neighboring relations, the number of handover, and handover success rate to a faraway neighbor are also other indications of over-shooter. This will lead to antenna down-tilting or reduce transmit power (Fig. 8.8).

The coverage hole is determined by cell layout, high number of poor RSRP, handover failure rate, drop calls and access failures or high number of MRO adjustment that does not improve those KPIs.

Once the KPIs are analyzed, CCO will determine whether overshooting or coverage hole is the dominate effect for evaluated cell based on its affected area. The affected area includes the cell, its neighbors, and its affected cells. The entire area will be frozen for the monitoring period to evaluate the impact of proposed change before the next run of change proposals. For this reason, CCO is lengthier processing than that of load balancing. It can change in days or weeks, instead of hours.

It is critical to have mechanism in place to safeguard the SON algorithm. SON will actively monitor segment where active changes are made, evaluate its effect on the network against predication. In case degradation or trendy to the wrong direction, SON could automatically revert back to previous parameter setting.

8.4 SON Deployment and Performance Benefits

SON functionality has been activated since LTE launch in 2010. However, hybrid SON is still in its infancy, in particular, coordination of distributed eNB function and 3rd party's centralized SON functionalities. Operator is playing an essential and more active rule to ensure smooth and coordinated operations of various SON features. We will discuss an example for potential hybrid SON deployment and some example and benefit from SON.

8.4.1 SON Deployment

Hybrid SON deployment required a coordinated control system at both eNB and OSS. Operators define a centralized policy management and SON server that include target key performance requirement and boundary of various parameters for various services.

The control system has mechanisms resided at both distributed eNB and centralized OSS with weighting and priorities of centralized and distributed features based on various relationships among cell sites, as illustrated in Fig. 8.9.

The control system creates an efficient SON operation based on operator operation policy and taking into account distance, path-loss, existing intra and inter-frequency neighbor relationships, manual operator biasing, restricted and default value, and other network characteristics, as illustrated in Fig. 8.9.

Operator will define policy with the distance and RF characteristics for each cell in the neighbor list of an eNB in a centralized database and creates a weighting factor (fc) based on those distance and RF characteristics. The RF characteristics could be associated with path-loss, received signal strength or SINR, network loading characteristic and inter-frequency path-loss discrepancies, operator biasing, restricted and default values, etc. Then, it creates an eNB parameter value based on historic and current proposed changes from the distributed and centralized algorithms:

$$P(t) = Pd(t) + g\{fc, Pc(t), Pd(t-1)\}$$

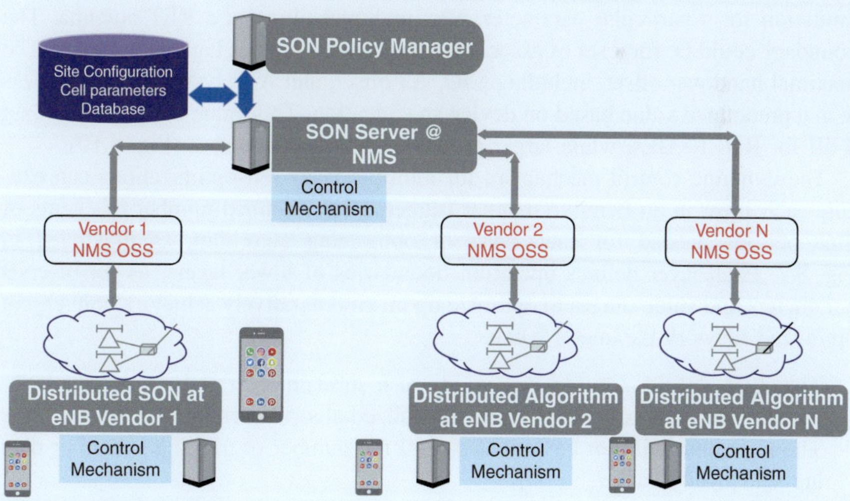

Fig. 8.9 SON deployment architecture

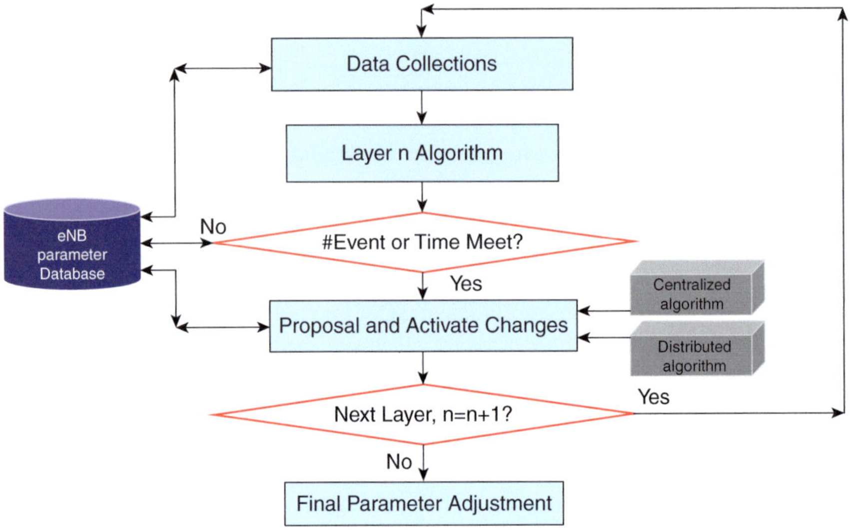

Fig. 8.10 SON policy control flowchart

where *P*(*t*) is the eNB parameter adjustment. *Pc*(*t*) and *Pd*(*t*) are the proposed change by centralized algorithm and eNB algorithm at time *t*, respectively.

Several algorithms could also try to optimize the same parameter. The final adjustment could be limited to a preset boundary. Those boundary could be the limitation for a particular parameter, e.g., maximal tilting at a RET antenna. The boundary could be for a set of associated parameters, such as handover values. The maximal handover offset, including CIO, A3 offset, and A3_Hysteresis, shall be less than a predefined value based on device specification. This value could be less than 9 dB for Rel. 10 UEs, while larger as UE sensitivities increases (Fig. 8.10).

The dynamic control mechanism for multiple layer self-optimizations can execute each layer in an iterative manner triggered by a defined number of events or time periods needed for each layer of optimizing algorithm, as illustrated in Fig. 8.7. Each layer defines operation boundaries of lower layers, as set of eNB parameters and value ranges to optimize upon. And iteratively achieve a convergent optimized network parameter value.

- The eNB algorithm could be running at a short interval and conducting optimization on an hourly basis, while centralized algorithm running at a daily basis
- The algorithm will not be activated until the number of trigging events or data are statistically sufficient.

Table 8.1 Example cell site database

Cell Site	Neighbor list	Distance mile	RF character path-loss from site (dB)	Weighting	A3 offset	Hysteresis	CIO Abs boundary	Acc. Offset boundary	Other
i-001	j-001	0.5	50	0.40	3	1	3	6	
	027	1.5	30	0.22	2	1	3	6	
	039	2	40	0.13	3	1	3	6	
j-001	i-001	0.5	60	0.33	3	1	3	6	
	045	20	100	0.01	3	1	3	8	
	067	30	120	0.00	2	3	3	8	
…		…	…		…	…	…		

8.4.2 Performance Benefits

SON benefits depending on network configuration, complexity, coordination of centralized and distributed SON, and underline network manual operation.

Table 8.1 and Fig. 8.5 illustrated a SON use case for a proposed combination of distributed and centralized SON algorithm in previous chapter. The weighting factor for each link is generated inverse to that of path-loss. The accumulated weighting factor of the centralized algorithm value on cell site i-001 and j-011 is 0.74 and 0.34, respectively. Therefore, the centralized algorithm is weighted higher in cell site i-001 than that of j-011. If the proposed change for both cell sites is 1 dB at eNB, and 2 dB at centralized algorithm, then, final parameter adjustment for cell site i-001 will be 2 dB, while cell site j-011 will only be 1 dB, both rounded to the nearest integer (Fig. 8.11).

SON servers could be distributed and co-located with OSS servers. This will ensure low latency in term of SON actions. Active load balancing and failover could be implemented to ensure robustness and efficiency. It can also centralized in an operator's network operation center (NOC) to ensure overall centralized control, sharing among various OSS servers and avoid SON segmentation.

SON can improve network operation efficiency and enable automation. At a sector level, SON could optimize antenna tilting and reduce drop call and drop call rate as illustrated in Fig. 8.12. On a cluster level, SON benefits vary depending on level of underline network optimization. It could improve call drop and throughput by around 10% in a well-optimized network to much large gain at an underperformed network. Other examples also include increase the traffics while maintain the same access success rate and drop call rate.

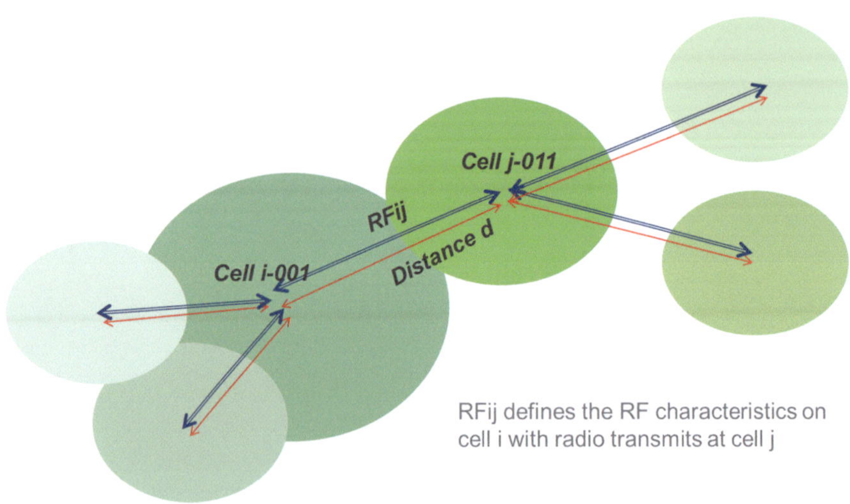

Fig. 8.11 SON use case example

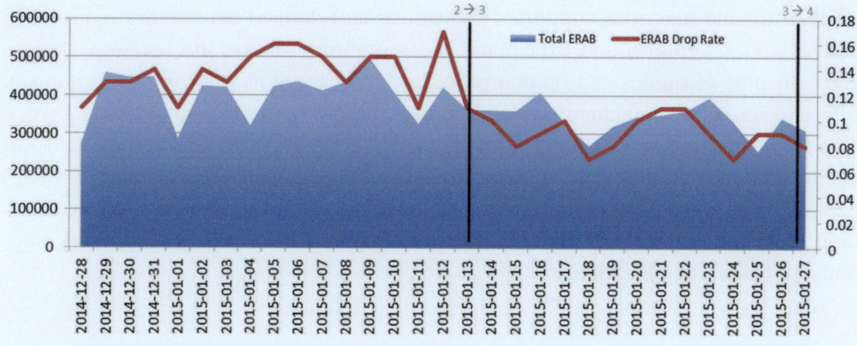

Fig. 8.12 Radio access bearer drop rate improvement

Benefits of SON demonstrated here are on radio network performance and optimization. As network performance fault analytics evolved to end-to-end service, we are working toward end-to-end SON optimization. This will be linked with programmable core network elements and service elements.

8.5 Evolved Network Architecture

SON, together with virtualization and software-defined networking, built the foundation for future architecture with automation and programmability. Those will drive network toward less complexity and low-cost network operation, fast deliveries of various services and business applications.

Figure 8.13 illustrates an architecture based on this principle. Business applications and servers' layer support various use cases of operators, enterprises, verticals, and third parties' application and servers. Network function and value

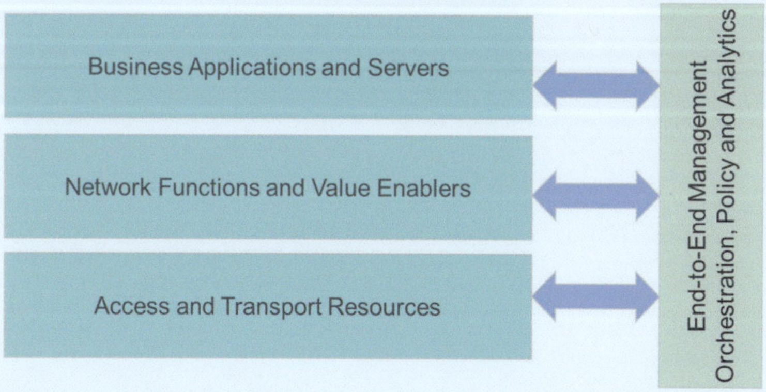

Fig. 8.13 Evolved network architecture

enabler include functions supported by various soft-defined networking modules to deliver various control and user plane configurations for specific services. Access and transport resources can be controlled properly by SON and other mechanism. The end-to-end management controls cross-domain operations of various layers. The orchestration enables automation across different types of access and transport resources and uses defined networking functions to provide desired network behavior and optimized configuration for each application. This network configuration and automation are aligned with and enabled by a policy framework supported by analytics insights.

8.6 Summary

SON is an enabler to build a smart infrastructure for change. We envision the SON will be first step and foundation to enable network robustness and flexibility and will let us to build a smart and intelligent network.

The policy and resource management at SON will get data not only from eNBs and UEs, but also from session, applications, and quality of service policies. Through more intensive analytics and decision engine, we will ensure a more robust and more reliable SON. Those real-time analytics will improve user experience and network efficiency.

Chapter 9
Summary

LTE wireless networks are composed of distinctly different network components in terms of protocols, architectures, network elements, and Network Management Systems. LTE wireless service providers have to deal with vastly different vendors and systems to manage and operate entire network. This book has described and proposed novel approaches leveraging new technologies, particularly in the area of Big Data Analytics to more efficiently manage and operate an LTE wireless network.

The proposed integrated solutions can effectively allow the LTE network service provider to operate entire integrated network, from RAN to Core, from UE to Application Service, as one unified system and correspondingly collect key metrics and data that are aligned, run through key analytics algorithms that considers all the metrics from each network and optimizes the network in terms of performance and fault.

9.1 LTE Wireless Network Element

The LTE Wireless Network Performance and Fault consists of the network performance and management of network elements in EUTRAN, EPC, and IP transport components, not only individually but also inter-working of these components. The key metrics for EUTRAN include radio access network accessibility, retainability, integrity, availability, and mobility. The key metrics for EPC include MME accessibility, mobility and capacity, SGW, PGW capacity, and connectivity. The key metrics for IP transport include latency, throughput, packet loss rate, and jitter. These metrics from different domains must be analyzed holistically to provide end-to-end view of network performance and fault analysis. For example, the key metric accessibility in itself is only meaningful in the EUTRAN domain and does not map to core network element. However, if accessibility is poor, it will degrade not only EUTRAN, but also overall performance, including the Core since the Core

D. Kakadia et al., *Network Performance and Fault Analytics for LTE Wireless Service Providers*, DOI 10.1007/978-81-322-3721-1_9

is dependent on EUTRAN and EPC. Also, the congestion and connection broken in core could lead to device fail to access the EUTRAN. In fact, they are all dependent on each other.

9.2 Analytics

Formulation of the best analytical techniques to solve LTE Wireless Network Performance and Fault problem statements is critical. In this book, we described fundamental analytics then built on that. We also further described how more advanced analytic techniques can be used to solve more complex wireless network problems which cannot be solved easily with basic analytics. We started with statistical process control (SPC) and built from there to outlier detection, regression, and forecasting. We also showed how Central Limit Theorem can be applied to reasonably simplify the problems being solved, as well as outlined why in some cases it does not help, and how more advanced methods can be used in such scenarios. We showed how to apply modern statistical tools (machine learning) to extracting useful information from the enormous amount of data on accessibility and other LTE Network KPIs. We also discussed how queuing theory can be used to explain dynamics of downlink and uplink throughput and other important measurements. Finally, we described methods that can be used in making predictions about data behavior based on previous behavior (forecasting) and business metrics (regression). Modern computing infrastructure makes applications of such techniques feasible and useful.

9.3 Solutions

This book described the main network segments—RAN, Backhaul, Metro, and Core of a typical LTE Wireless Service Provider Network. The book then described the network equipment that makes up these network segments and LTE functions: enodeB, SGW, PGW, Routers, ROADMS, etc. This provides the reader with a good background of the LTE service provider problem domain.

New solutions to LTE Service Network Performance and Fault Management systems using Big Data Analytics specific to LTE were proposed. These new approaches described the limitations of currently available systems and overcame those challenges by using advanced analytics to provide unprecedented insights which can be automated that significantly improve operational efficiencies and provide the same or better quality of experience to subscribers by making better use of existing resources by tuning parameters based on analytics described in this book.

Self-organized networks architecture is presented as a way to utilize network performance and fault analytics to enable network automation. SON can significantly improve operational efficiencies and speed up network deployment. Hybrid SON architecture with both distributed functionality at eNB and centralized functionality at NMS can further improve network performance. Automate Neighbor Relations, Mobility Robustness Optimization, Mobile Load Balancing, Coverage, and Capacity Optimization are essential components of self-network optimization. They work effectively together to improve network accessibility, retainability, throughput, latency, and mobility. SON will also adjust network configuration per service type according to end-to-end performance outlined in this book to improve overall user experience in future.

Big Data Analytics proposed in this book is still at its infancy. It will be specifically used to improve overall network performance and service quality.

References

[MIMO2011] MIMO Performance at 700 MHz: Field Trials of LTE with Handheld UE. In: Proceedings of VTC fall 2011, 5 Sept 2011

[3GPTR905] 3GPP TR 21.905: Vocabulary for 3GPP Specifications

[3GPPTS22] 3GPP TS 32.522: Telecommunication management; Self-Organizing Networks (SON) Policy Network Resource Model (NRM) Integration Reference Point (IRP); Information Service (IS)

[3GPPTS500] 3GPP TS 32.500: Telecommunication Management; Self-Organizing Networks (SON); Concepts and requirements

[3GPTS501] 3GPP TS 32.501: Telecommunication management; Self-Configuration of Network Elements; Concepts and Integration Reference Point (IRP) Requirements

[3GPPTS511] 3GPP TS 32.511: Telecommunication management; Automatic Neighbour Relation (ANR) management; Concepts and requirements

[3GPPTS521] 3GPP TS 32.521: Telecommunication management; Self-Organizing Networks (SON) Policy Network Resource Model (NRM) Integration Reference Point (IRP); Requirements

[3GPPTS541] 3GPP TS 32.541: Telecommunications Management; Self-Organizing Networks (SON); Self-healing concepts and requirements

[3GPPTS401] 3GPP TS 32.401: Telecommunication management; Performance Management (PM); Concept and requirements

[3GPPTS600] 3GPP TS 32.600: Telecommunication management; Configuration Management (CM); Concept and high-level requirements

[3GPPTS301] 3GPP TS 24.301: Technical Specification Group Core Network and Terminals; Non-Access-Stratum (NAS) protocol for Evolved Packet System (EPS); Stage 3

[3GPPTS274] 3GPP TS 29.274: Evolved General Packet Radio Service (GPRS); Tunnelling Protocol for Control plane (GTPv2-C); Stage 3

[3GPPTS4018] 3GPP TS 23.401: General Packet Radio Service (GPRS) enhancements for Evolved Universal Terrestrial Radio Access Network (E-UTRAN) access (Release 8)

[3GPPTS74] 3GPP TS 29.274: Tunnelling Protocol for Control plane (GTPv2-C)

[3GPPTS281] 3GPP TS 29.281: GPRS Tunnelling Protocol User Plane (GTPv1-U)

[3GPPTS414] 3GPP TS 36.414: Evolved Universal Terrestrial Access Network (E-UTRAN); S1 data transport

[3GPPTS272] 3GPP TS 29.272: Mobility Management Entity (MME) and Serving GPRS Support Node (SGSN) related interfaces based on Diameter protocol
[3GPPTS203] 3GPP TS 23.203: Policy and charging control architecture
[3GPPTS13] 3GPP TS 36.413: Evolved Universal Terrestrial Radio Access Network (E-UTRAN); S1 Application Protocol (S1AP)
[3GPPTS412] 3GPP TS 36.412: Evolved Universal Terrestrial Access Network (E-UTRAN); S1 signalling transport
[3GPPTS402] 3GPP TS 23.402: Architecture enhancements for non-3GPP accesses
[IETFRFC5136] IETF RFC 5136: Defining Network Capacity
[3GPPTS246] 3GPP TS 23.246: Multimedia Broadcast Multicast Service (MBMS); Architecture and functional description
[3GPPTS445] 3GPP TS 36.445: Evolved Universal Terrestrial Radio Access Network (E-UTRAN); M1 data transport
[3GPPTS272] 3GPP TS 23.272: Circuit Switched (CS) fallback in Evolved Packet System (EPS); Stage 2
[3GPPTS212] 3GPP TS 29.212: Policy and charging control over Gx reference point
[3GPPTS213] 3GPP TS 29.213: Policy and charging control signalling flows and Quality of Service (QoS) parameter mapping
[IETFRFC88] IETF RFC 3588: Diameter Base Protocol
[KPR2015] Kakadia D, Parker BJ, Reyes EMD (2015) Performance and fault management system. United States Patent 8972798
[KLT2014] Kakadia D, Lee J, Tan T (2014) TCP flow control optimized for networks having radio segments. United States Patent 8885476
[KL2014] Kakadia D, Lopez JM (2014) Optimizing performance information collection. United States Patent 8627468
[KLT2013] Kakadia D, Lee JJ, Tan TH (2013) System and method for multi-layer network analysis and design. United States Patent 8369222
[SCRC2004] Scrucca L (2004) qcc: an R package for quality control charting and statistical process control. R News 4/1:11-17
[GILG2010(1)] Gilgur A, Perka M (2010) Computer storage capacity forecasting system using cluster-based seasonality analysis. US Patent 7783510. Filed 06/25/2007. Awarded 08/24/2010
[GILG2014] Gilgur A, Ferrandiz J (2014) Predictive SPC—presented at the conference on statistical practice (CSP2014). American Statistical Association, Tampa, FL, Feb 2014
[BRTN2012] Barton T (2015) Cluster analysis of time series data. Downloaded May 30, 2015 from http://ai.ms.mff.cuni.cz/~sui/barton.pdf
[TOOM2014] Toomey D (2014) R for data science PACKT publishing. ISBN 9781784390860
[JSHI2014] Shi J, Mamoulis N, Wu D, Cheung DW (2014) Density-based place clustering in geo-social networks
[TWRS2013] Polymatheia Blog. Posted on 24 Oct 2013 by Sherry Towers http://sherrytowers.com/2013/10/24/k-means-clustering/
[HNNG2014] Hennig C (2014) fpc: flexible procedures for clustering. R package version 2.1-9. http://CRAN.R-project.org/package=fpc

MIX
Papier aus verantwortungsvollen Quellen
Paper from responsible sources
FSC® C105338

If you have any concerns about our products,
you can contact us on
ProductSafety@springernature.com

In case Publisher is established outside the EU,
the EU authorized representative is:
Springer Nature Customer Service Center GmbH
Europaplatz 3, 69115 Heidelberg, Germany

Printed by Libri Plureos GmbH
in Hamburg, Germany